METHODS IN MOLECULAR BIOLOGY

Series Editor
John M. Walker
School of Life and Medical Sciences
University of Hertfordshire
Hatfield, Hertfordshire, UK

For further volumes:
http://www.springer.com/series/7651

For over 35 years, biological scientists have come to rely on the research protocols and methodologies in the critically acclaimed *Methods in Molecular Biology* series. The series was the first to introduce the step-by-step protocols approach that has become the standard in all biomedical protocol publishing. Each protocol is provided in readily-reproducible step-by-step fashion, opening with an introductory overview, a list of the materials and reagents needed to complete the experiment, and followed by a detailed procedure that is supported with a helpful notes section offering tips and tricks of the trade as well as troubleshooting advice. These hallmark features were introduced by series editor Dr. John Walker and constitute the key ingredient in each and every volume of the *Methods in Molecular Biology* series. Tested and trusted, comprehensive and reliable, all protocols from the series are indexed in PubMed.

Clostridioides difficile

Methods and Protocols

Edited by

Kurt H. Piepenbrink

Department of Food Science and Technology, University of Nebraska-Lincoln, Lincoln, NE, USA; Department of Biochemistry, University of Nebraska-Lincoln, Lincoln, NE, USA; Department of Chemistry, University of Nebraska-Lincoln, Lincoln, NE, USA; Nebraska Food for Health Center, University of Nebraska-Lincoln, Lincoln, NE, USA; Center for Integrated Biomolecular Communication, University of Nebraska-Lincoln, Lincoln, NE, USA

Jennifer M. Auchtung

Department of Food Science and Technology, University of Nebraska-Lincoln, Lincoln, NE, USA; Nebraska Food for Health Center, University of Nebraska-Lincoln, Lincoln, NE, USA

Humana Press

Editors
Kurt H. Piepenbrink
Department of Food Science and Technology
University of Nebraska-Lincoln
Lincoln, NE, USA

Department of Biochemistry
University of Nebraska-Lincoln
Lincoln, NE, USA

Department of Chemistry
University of Nebraska-Lincoln
Lincoln, NE, USA

Nebraska Food for Health Center
University of Nebraska-Lincoln
Lincoln, NE, USA

Center for Integrated Biomolecular Communication
University of Nebraska-Lincoln
Lincoln, NE, USA

Jennifer M. Auchtung
Department of Food Science and Technology
University of Nebraska-Lincoln
Lincoln, NE, USA

Nebraska Food for Health Center
University of Nebraska-Lincoln
Lincoln, NE, USA

ISSN 1064-3745 ISSN 1940-6029 (electronic)
Methods in Molecular Biology
ISBN 978-1-0716-5327-2 ISBN 978-1-0716-5328-9 (eBook)
https://doi.org/10.1007/978-1-0716-5328-9

This Humana imprint is published by the registered company Springer Science+Business Media, LLC, part of Springer Nature.
The registered company address is: 1 New York Plaza, New York, NY 10004, U.S.A.

Preface

This volume is an update to another MIMB volume titled *Clostridium difficile: Methods and Protocols* that was published in 2016. Since 2016, we have gained a greater understanding of *C. difficile*, including an appreciation that the genus name "*Clostriodiodes*" better reflects the phylogeny of *C. difficile* than its previous genus name, which was initially assigned based on its physiology as an anaerobic, spore-forming pathogen. There have also been numerous updates to the approaches used to characterize *C. difficile* physiology, including new methods for genetic modifications and novel models for studying host-pathogen interactions. Gaining a deeper understanding of *C. difficile* physiology is crucial for developing new therapies for treatment and predicting the future evolution of this important clinical pathogen. While the detailed protocols shared in this edition represent new approaches for characterizing *C. difficile*, they also build upon a previous foundation of well-established protocols described in the first edition of this series, which is still being used today.

Lincoln, NE, USA

Kurt H. Piepenbrink
Jennifer M. Auchtung

Contents

Contributors

Carolina Altaf Figueiredo • *Institut Pasteur, Université Paris-Cité, UMR-CNRS 6047, Laboratoire Pathogenèse des Bactéries Anaérobies, Paris, France*

Jennifer M. Auchtung • *Department of Food Science and Technology, University of Nebraska-Lincoln, Lincoln, NE, USA; Nebraska Food for Health Center, University of Nebraska-Lincoln, Lincoln, NE, USA*

Baishakhi Biswas • *Department of Food Science and Technology, University of Nebraska-Lincoln, Lincoln, NE, USA; Nebraska Food for Health Center, University of Nebraska-Lincoln, Lincoln, NE, USA*

Joshua N. Brehm • *Department of Biology, Texas A&M University, College Station, TX, USA*

Robert A. Britton • *Department of Molecular Virology and Microbiology, Baylor College of Medicine, Houston, TX, USA; Alkek Center for Metagenomics and Microbiome Research, Baylor College of Medicine, Houston, TX, USA*

Thomas Candela • *Micalis Institute, Université Paris-Saclay, INRAE, AgroParisTech, Jouy-en-Josas, France*

Léo C. Caulat • *Institut Pasteur, Université Paris Cité, Laboratoire Pathogenèse des Bacté ries Anaérobies, Paris, France; Bacterial Systems Biology and Anti Microbial Resistance, Laboratoire de Biologie et Pharmacologie Appliquée, École Normale Supérieure Paris-Saclay, Gif-sur-Yvette, France*

Erin C. Chard • *Department of Regenerative Medicine & Cell Biology, Medical University of South Carolina, Charleston, SC, USA*

Ian Cheong • *Temasek Life Sciences Laboratory, Singapore, Singapore; Department of Biological Sciences, National University of Singapore, Singapore, Singapore*

Tsz Ting Chow • *Prismea Limited, London, UK*

James Collins • *Department of Microbiology & Immunology, University of Louisville, Louisville, KY, USA; Center for Predictive Medicine, University of Louisville, Louisville, KY, USA; Center for Microbiomics, Inflammation and Pathogenicity, University of Louisville, Louisville, KY, USA*

Jeroen Corver • *Leiden University Center for Infectious Diseases, Leiden University Medical Center, Leiden, The Netherlands*

Deiziane V. S. Costa • *Division of Infectious Diseases and International Health, University of Virginia, Charlottesville, VA, USA*

Eliane de Oliveira Ferreira • *Laboratory of Biology of Anaerobes, Medical Microbiology Department, Paulo de Goes Microbiology Institute, Federal University of Rio de Janeiro (UFRJ), Rio de Janiero, Brazil*

Nicholas V. DiBenedetto • *Department of Molecular Biology and Microbiology, Tufts University School of Medicine, Boston, MA, USA; Program in Molecular Microbiology, Tufts University Graduate School of Biomedical Sciences, Boston, MA, USA*

Bruno Dupuy • *Institut Pasteur, Université Paris-Cité, UMR-CNRS 6047, Laboratoire Pathogenèse des Bactéries Anaérobies, Paris, France*

Melinda A. Engevik • *Department of Regenerative Medicine & Cell Biology, Medical University of South Carolina, Charleston, SC, USA; Department of Pharmacology & Immunology, Medical University of South Carolina, Charleston, SC, USA*

DANIEL ERICKSON • *Department of Microbiology & Immunology, University of Louisville, Louisville, KY, USA; Center for Predictive Medicine, University of Louisville, Louisville, KY, USA*

LOUIS-CHARLES FORTIER • *Department of Microbiology and Infectious Diseases, Faculty of Medicine and Health Sciences, Université de Sherbrooke, Sherbrooke, Québec, Canada*

AURÉLIE LOTOUX • *Institut Pasteur, Université Paris Cité, Laboratoire Pathogenèse des Bacté ries Anaérobies, Paris, France*

JEANNE MALET-VILLEMAGNE • *Micalis Institute, Université Paris-Saclay, INRAE, AgroParisTech, Jouy-en-Josas, France*

ISABELLE MARTIN-VERSTRAETE • *Institut Pasteur, Université Paris Cité, Laboratoire Pathogenèse des Bactéries Anaérobies, Paris, France; Institut Universitaire de France, Paris, France*

HUGH C. MCCULLOUGH • *Department of Food Science and Technology and Nebraska Food for Health Center, University of Nebraska-Lincoln, Lincoln, NE, USA; Department of Biosystems Engineering, University of Nebraska-Lincoln, Lincoln, NE, USA*

CLAIRE MORVAN • *Institut Pasteur, Université Paris Cité, Laboratoire Pathogenèse des Bacté ries Anaérobies, Paris, France*

ANA M. OLIVEIRA PAIVA • *Leiden University Center for Infectious Diseases, Leiden University Medical Center, Leiden, The Netherlands*

ASSILINA PARFUT • *Micalis Institute, Université Paris-Saclay, INRAE, AgroParisTech, Jouy-en-Josas, France; Plateau technique de Microbiologie, CHU de Strasbourg, Strasbourg, France*

KURT H. PIEPENBRINK • *Department of Food Science and Technology, University of Nebraska-Lincoln, Lincoln, NE, USA; Nebraska Food for Health Center, University of Nebraska-Lincoln, Lincoln, NE, USA; Department of Biochemistry, University of Nebraska-Lincoln, Lincoln, NE, USA; Department of Chemistry, University of Nebraska-Lincoln, Lincoln, NE, USA; Center for Integrated Biomolecular Communication, University of Nebraska-Lincoln, Lincoln, NE, USA*

JASON PIZZINI • *Department of Molecular Virology and Microbiology, Baylor College of Medicine, Houston, TX, USA*

MARK D. PRESTON • *Prismea Limited, London, UK*

KATE PSENKA • *Department of Regenerative Medicine & Cell Biology, Medical University of South Carolina, Charleston, SC, USA*

ALEXIA L. M. ROYER • *Department of Microbiology and Infectious Diseases, Faculty of Medicine and Health Sciences, Université de Sherbrooke, Sherbrooke, Québec, Canada; Université Paris-Saclay, CEA, CNRS, Institute for Integrative Biology of the Cell (I2BC), Gif-sur-Yvette, France*

AIMEE SHEN • *Department of Molecular Biology and Microbiology, Tufts University School of Medicine, Boston, MA, USA*

BEN S. SIDNER • *Department of Food Science and Technology, University of Nebraska-Lincoln, Lincoln, NE, USA; Nebraska Food for Health Center, University of Nebraska-Lincoln, Lincoln, NE, USA*

WIEP KLAAS SMITS • *Leiden University Center for Infectious Diseases, Leiden University Medical Center, Leiden, The Netherlands*

JOSEPH A. SORG • *Department of Biology, Texas A&M University, College Station, TX, USA*

RICHARD A. STABLER • *Department of Infection Biology, London School of Hygiene and Tropical Medicine, London, UK*

XINGMIN SUN • *Department of Molecular Medicine, Morsani College of Medicine, University of South Florida, Tampa, FL, USA*
AJITHA SUNDARESAN • *Temasek Life Sciences Laboratory, Singapore, Singapore*
ANNA M. TINGLER • *Department of Regenerative Medicine & Cell Biology, Medical University of South Carolina, Charleston, SC, USA*
ANDREW A. UMANSKY • *Department of Microbiology and Infectious Diseases, Faculty of Medicine and Health Sciences, Université de Sherbrooke, Sherbrooke, Québec, Canada*
FRANCISCO A. UZAL • *California Animal Health and Food Safety Laboratory-San Bernardino Branch, School of Veterinary Medicine, University of California-Davis, San Bernardino, CA, USA*
SHAOHUI WANG • *Department of Molecular Medicine, Morsani College of Medicine, University of South Florida, Tampa, FL, USA*
CIRLE A. WARREN • *Division of Infectious Diseases and International Health, University of Virginia, Charlottesville, VA, USA*
DUOLONG ZHU • *Department of Molecular Virology and Microbiology, Baylor College of Medicine, Houston, TX, USA; Alkek Center for Metagenomics and Microbiome Research, Baylor College of Medicine, Houston, TX, USA*
MARIE-EMELINE ZIELINSKI • *Micalis Institute, Université Paris-Saclay, INRAE, AgroParisTech, Jouy-en-Josas, France; Hôpital de la Pitié Salpêtrière, Laboratoire de virologie, Paris, France*

Chapter 1

From FASTQ to Function to Phylogeny: In Silico Methods for Antimicrobial Resistance and Outbreak Investigation

Mark D. Preston, Tsz Ting Chow, and Richard A. Stabler

Abstract

This chapter presents a method to process *Clostridioides difficile* whole genome next-generation sequencing (WGS) data for draft genome assembly, gene annotation, multilocus strain type identification (MLST), antimicrobial resistance (AMR), and generation of a relationship phylogeny. We use publicly available data to do a mock outbreak analysis to demonstrate how WGS can provide in-depth insight into these isolates.

Key words Whole genome sequencing, Gene annotation, MLST, Outbreak analysis, *C. difficile*, Bioinformatics

1 Introduction

Bioinformatic analyses of bacterial sequencing data are now often able to be completed on laptop computers, enabling the laboratory scientist the opportunity to perform their own analyses. These analyses are composed of multiple bioinformatics tools (a pipeline) applied in a common order to consistently reach robust conclusions. The individual software tools themselves are open source, generally very well established, and do not require deep understanding of the underlying algorithms to use well.

This chapter demonstrates how to process individual bacterial isolates to a draft genome which can then be used to characterize each genome (mutations, gene annotation, MLST) and to characterize sets of genomes (core and accessory genes, core mutations, potential outbreaks). We start with Illumina next-generation sequencing (NGS) data for each sample in the form of FASTQ data files and at each step describe outputs (tables, graphs, spreadsheets, etc.) that give insight and enable troubleshooting and monitoring of the pipeline.

Kurt H. Piepenbrink and Jennifer M. Auchtung (eds.), *Clostridioides difficile: Methods and Protocols*, Methods in Molecular Biology, vol. 3046, https://doi.org/10.1007/978-1-0716-5328-9_1,

Outbreak Analysis For our description of outbreak analysis, a subset of data from "*Reference Susceptibility Testing and Genomic Surveillance of Clostridioides difficile, United States, 2012–17*" [1] will be used. Information on the selected isolates are in the metadata files. Some of the data is not from the original study but has been added for the purposes of this chapter. In this chapter we will be using the data from the isolates to investigate if there is an outbreak of *C. difficile*, including which clones are linked with antimicrobial resistance.

2 Materials

2.1 Computer Setup

Most bioinformatics tools are designed to run on the Linux command line. This protocol assumes that you have access to a Linux system, either directly via a Linux computer, using the Terminal App on a Mac, or via the Windows Subsystem for Linux (WSL) on a Windows computer. WSL is an optional Windows component; see [2] on instructions for activating.

This example pipeline uses 15+ bioinformatics tools and NGS data for 16 samples and these need to be placed on your computer system.

Each bioinformatics software tool can depend on many other software tools and libraries to be installed to run correctly. The 15+ tools that we will use in this protocol results in a *dependency tree* of over 100 pieces of software. Fortunately, specialist tools have been developed to manage the installation of bioinformatics software and their dependencies. One of the most popular is called conda [3].

We will require other, non-bioinformatic tools too such as an unzip program (`unzip`), a program to access online code repositories (`git`), to download data files from the Internet (`wget`) and to manipulate data (`sed`). Please refer to your specific Linux installation methods, but on an Ubuntu-based system the commands to install these are:

```
$ sudo apt-get update
$ sudo apt-get install git sed unzip wget
```

Linux Box

This chapter presumes the reader has access to a Linux machine and has working knowledge of the Linux operating system. Linux commands are indicated with a "$" and written

(continued)

in `Courier new` font, the "\" at the end of command lines is a textual indicator to not press return but to keep on typing and without typing the "\".

The command lines can use two special characters. ">" will *redirect* the normal tool output from writing to the screen, to writing to a file instead—allowing output to be saved to disk easily. The "*" can be used as a *wildcard* to select multiple files at once—so that all samples can be run with one command; see `abricate`, `roary`, and `chewBACCA` sections.

Installing software normally requires administrator privileges, and prefixing commands with sudo can achieve this. Do not use `sudo` for anything else such as data processing, as this can have unintended and disastrous consequences.

Most tools can take advantage of any parallel capabilities of your computer (called threads or cpus). We have assumed four threads throughout, but this can be changed to suit your machine. However, choosing the maximum available may cause memory or other processing problems. Tools take a parameter on the command line to instruct them to do this. We will see the parameter appearing in seven different ways below (`-t 4`, `--thread 4`, `--threads 4`, `-c 4`, `--cpu 4`, `--cpus 4`, and `-p 4`) with the choice of parameter name made by the tool's author.

2.2 Software Tools

The primary bioinformatics tools we will use are:

1. fastqc Reporting Analyses fastq data for quality, adapter sequences
2. fastp QC Removes low-quality reads from fastq data
3. unicycler Sequence creation De novo assembles sequences (contigs) from reads
4. quast Reporting Analyses assembly quality
5. abacas Sequence ordering By aligning contigs against a reference genome
6. prokka Gene annotation Transfer genes, CDS, RNA, etc.
7. mlst Sequence typing Compare gene combos to a database for typing
8. abricate Drug resistance genes Compare individual genes to a database for DR

9. snippy Core mutations Haploid variant calling and core genome alignment
10. fastree Phylogenic tree Use the core mutations to identify relationships
11. roary Core genome By identifying orthologous genes
12. chewBBACA Sequence typing Allelic typing approach across the core genome

Example code for this chapter is stored on an Internet code repository—github. The command to download the code is:

```
$        git clone https://github.com/markdpreston/cdiff-
chapter.git
```

Executing this command will have copied our example code to your computer. Most of the files are bash scripts: small programs that take parameters from the command line and run bioinformatics tools. The other two files are text files that store useful information (`sample_srr_numbers.txt`, `environment.yaml`).

To install all the bioinformatics tools, software, and libraries we recommend using conda. Conda creates an enclosed environment within, but without compromising, your Linux system. Conda has to be installed manually by downloading a script using wget and then running that script to install it.

```
$ wget https://repo.anaconda.com/miniconda/Miniconda3-latest-
Linux-x86_64.sh
$ bash ./Miniconda3-latest-Linux-x86_64.sh
```

During installation, conda will prompt you to review the license agreement and choose installation options. In particular, the installer will ask whether you want to initialize conda—it is recommended to answer "yes." After installation, close and reopen your terminal to activate the initialization. If successful, you will see the `(base)` environment name appear in your terminal prompt, indicating that conda has been initialized and is active.

Next, conda must be told where the bioinformatics tools are located on the Internet (bioconda). The list of bioinformatics tools for this chapter (`environment.yaml`), with their dependencies, is installed in an environment named (-n) cdiff. The final command activates this environment and makes the software available to us:

```
$ conda config --add channels bioconda
$ conda config --add channels conda-forge
$ conda env create --file environment.yaml
$ conda activate cdiff
```

After activating the cdiff environment, the prompt in your terminal will update to show `(cdiff)`, indicating the environment is active. If you have a need to close the Linux/WSL at any point, you should be able to resume your analysis. However, you will need to reenter the conda environment first:

```
$ conda activate cdiff
```

2.3 Download Data

We will use a subset of the publicly available WGS data from the Emerging Infections Program *C. difficile* Infection Working Group publication [1], with the *C. difficile* PCR-ribotype 027 (RT027) reference sequence R20291. All isolates are indistinguishable as RT027 samples and will be used to demonstrate how WGS data could be useful in an outbreak. To download the example raw NGS data, run the command:

```
$ ./download_data.sh data
```

This script contains the commands to download data directly from the NCBI using SRI accession numbers. The script uses the `prefetch` and `fasterq` bioinformatic tools augmented by standard Linux command line tools (`mkdir`, `for`, `gzip`) to download 16 samples and place them in the new `data` folder.

In short, the *C. difficile* DNA was extracted and broken into fragments approximately 1 kbp long. These fragments were then sequenced from both ends with an unsequenced gap in the middle. The length of the read is set by the Illumina machine during the sequencing read setup; in this case, the reads are 251 bp long. The data from the sequenced ends are known as the forward and reverse reads denoted by "_1" and "_2" in the file names, respectively. Together, they are known as paired-end reads in that the first sequencing *read* in the forward file is from the same DNA fragment as the first *read* from the reverse file.

In the following methods we refer to a generic sample (named `sample1`) so throughout each step any of the downloaded samples can be substituted in (e.g., replacing the text `sample1` with `SRR12437196`). We create our dummy sample1 by copying a real sample and use this to generate the results and images in our examples:

```
$ cp data/SRR12437196_1.fastq.gz data/sample1_1.fastq.gz
$ cp data/SRR12437196_2.fastq.gz data/sample1_2.fastq.gz
```

3 Methods

3.1 Reference Files and Preprocessing

In this chapter publicly available *C. difficile* sequence data will be used as an example but alternative data can be used. The reference sequences and preprocessing are performed using the command:

```
$ ./download_references.sh
```

Inside this script, the following actions occur: Firstly, it downloads the reference with annotation (.embl) and reference sequence only files (.fasta):

```
$ wget "https://www.ebi.ac.uk/ena/browser/api/embl/
FN545816.1?download=true" \
     -O ref/R20291.embl"
$ wget "https://www.ebi.ac.uk/ena/browser/api/fasta/
FN545816.1?download=true" \
     -O ref/R20291.fasta"
```

This will place the R20291 reference files in the new `ref` folder.

3.2 Quality Control: fastqc and fastp

Before attempting to assemble a genome from the raw data, it is advisable to remove poor-quality data that can have a negative impact on the downstream processing and final output. Fastq data not only contains sequence data but also an alphanumeric quality score (Phred score) per base, which represents the confidence level of a base call in the DNA sequence. We use this data to remove low-quality bases from the ends of reads (shortening them), to remove poor-quality reads (reads that have too many low-quality bases), and to remove contamination (such as from adapters added during the Illumina sequencing library preparation process).

The `fastqc` [4] program analyses the files taken from the sequencer. It creates text, image, and HTML report files that should be checked to ensure high-quality data entering the pipeline. We will run `fastqc` before and after removing unwanted data using `fastp` [5].

We use `fastqc` and specify three parameters—the number of CPU threads (-t, the more processors that your computer has, then the more processing can be performed in parallel and reduce the runtime), the individual fastq file to process, and the output folder (`fastqc_pre`) to write the results:

```
$      fastqc -t 4 data/sample1_1.fastq.gz -o fastqc_pre
```

By opening the file `fastqc_pre/sample1_1_fastqc.html` in a browser, you will see a graphical report on the sequencing quality of that file.

Among other data, the graphical report shows the average base quality per cycle/position, on a logarithmic scale (10 = 1 error in 10, 20 = 1 error in 100, 30 = 1 error in 1000 and 40 = 1 error in 10,000). The quality of the called bases (sometimes referred to as positions or cycles) drops off (see Fig. 1) and becomes more variable along the read; this is especially noticeable in the reverse reads (see Fig. 1b, d). Any quality score below 20 is deemed to be poor-quality data. Although 1 error in 100 (score of 20) seems very low, in the context of ~7.0×10^5 paired 251-bp reads yielding ~1.6×10^8 individual base calls, an error rate of 1 in 100 would potentially yield (in the order of) 10^6 erroneous calls. The other graphs in `fastqc` show more quality statistics related to the sample.

The low-quality data is removed using the `fastp` tool, which is called like this:

```
$ fastp --thread 4 -i data/sample1_1.fastq.gz -I data/sam-
ple1_2.fastq.gz \
     -o trim/sample1.p1.fastq.gz -O trim/sample1.p2.fastq.gz \
      --unpaired1 trim/sample1.s1.fastq.gz --unpaired2 trim/
sample1.s2.fastq.gz \
      --cut_right --cut_window_size 4 --cut_mean_quality 20 --
length_required 50 \
      --html trim/sample1.html --json trim/sample1.json
```

This `fastp` command line assesses the input files (`-i`, `-I`) read-by-read using a rolling base window of length 4 (`--cut_window_size`) and removes anything to the right (`--cut_right`) whenever the window's average base q-score drops below 20 (`--cut_mean_quality`) and outputs the resultant sequencing reads (`-o`, `-O`) if their length is above 50 (`--length_required`). In most cases, the paired reads remain paired, but in some cases one of the paired reads is removed completely leaving a read without a mate, i.e., unpaired. These unpaired reads are stored separately from the surviving paired reads (`--unpaired1`, `--unpaired2`). Note that `-i`, `-o`, and `--unpaired1` refer to the forward reads, and `-I`, `-O`, and `--unpaired2` refer to the reverse reads. Reports on bases and reads removed, insert size estimation, before and after filtering statistics, etc.

We run the `fastqc` command again:

```
$      fastqc -t 4 trim/sample1.p1.fastq.gz -o fastqc_post
```

And compare the results (`fastqc_post/sample1.p1_fastqc.html`) to the pre-QC `fastqc_pre` for the same sample's forward reads (compare Fig. 1a–c, b–d).

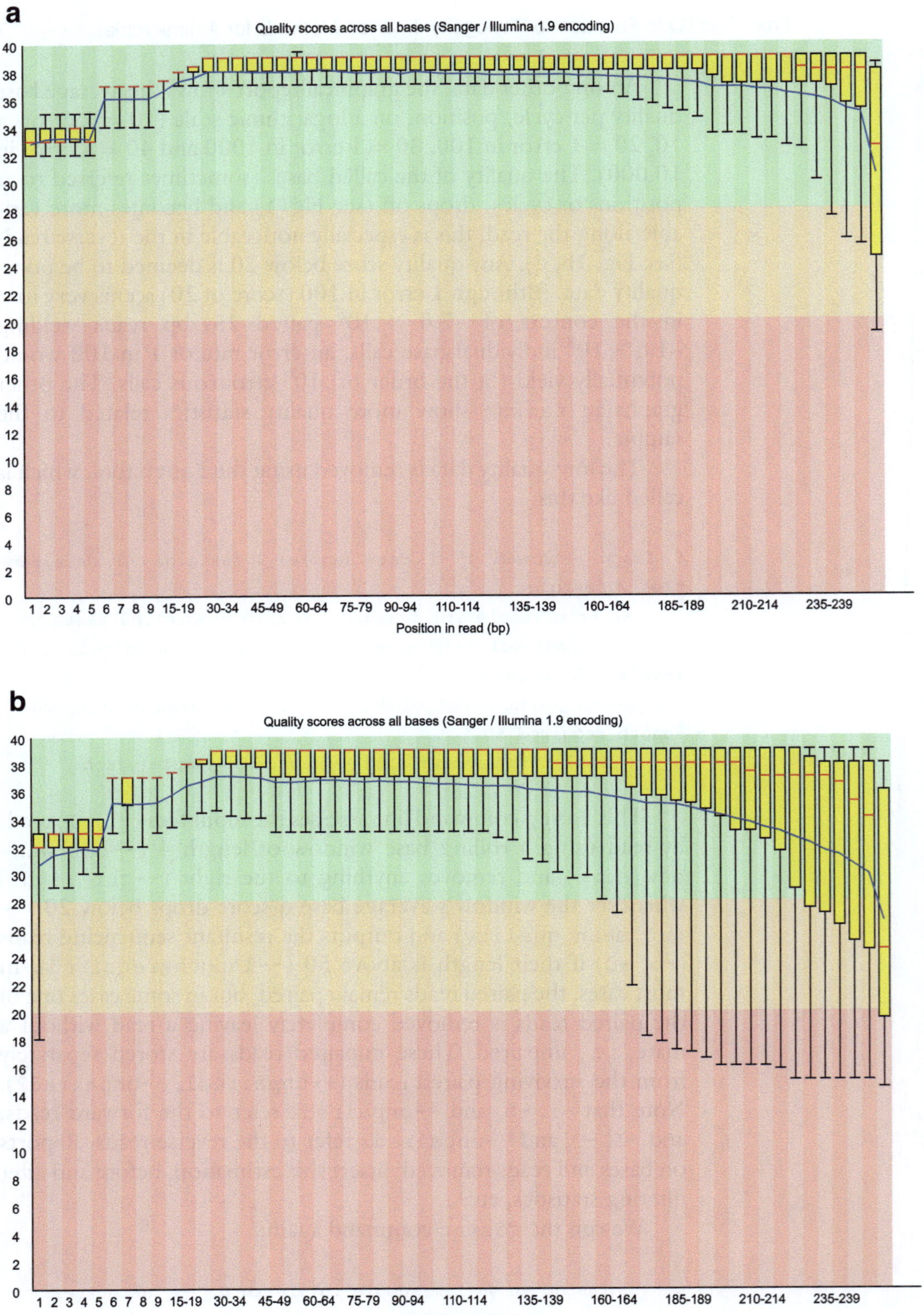

Fig. 1 Improvement of overall data quality by using a read trimmer. FASTQC "per base sequence quality" BoxWhisker plot summaries of all reads in paired fastq files. The BoxWhisker plots show the median value, the interquartile range (25–75%), the 10% and 90% points, and the mean quality averaged along reads. The phred scores (y-axis) indicate high-quality data (28–40, green), reasonable quality (20–28, orange), and poor quality (0–20, red). The y-axis indicates position within the reads from base 1 (first column) to final base (last column), with some positions being amalgamated (e.g., positions 235–239). (**a**) Forward reads before trimming, (**b**) reverse reads before trimming, (**c**) forward reads after trimming, and (**d**) reverse reads after trimming

c

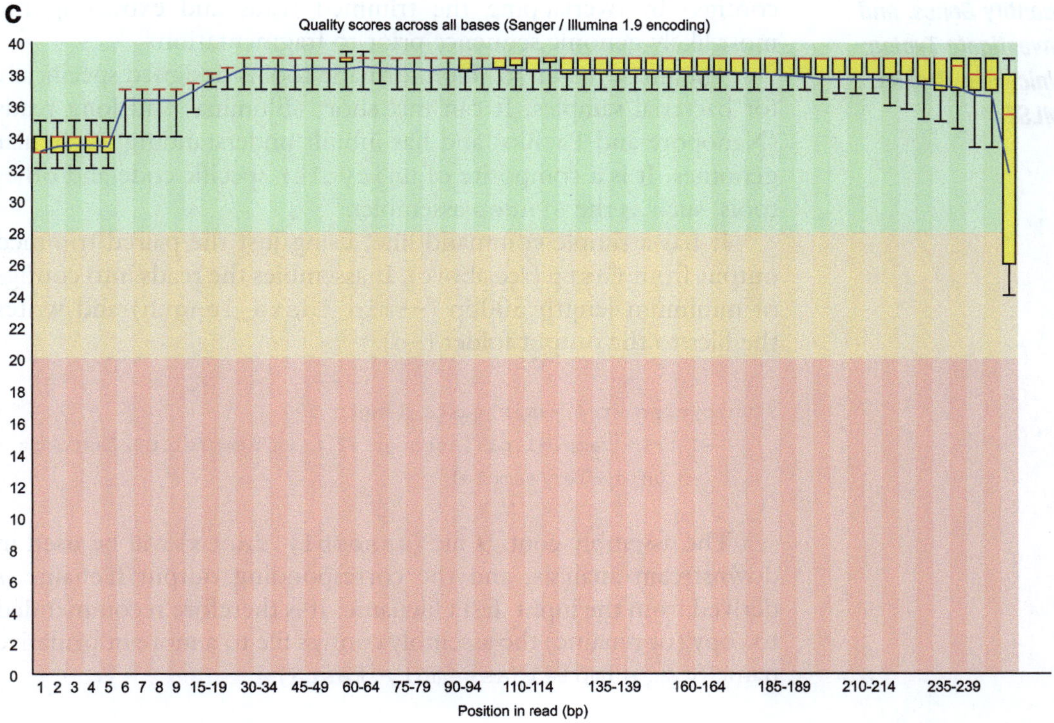

d

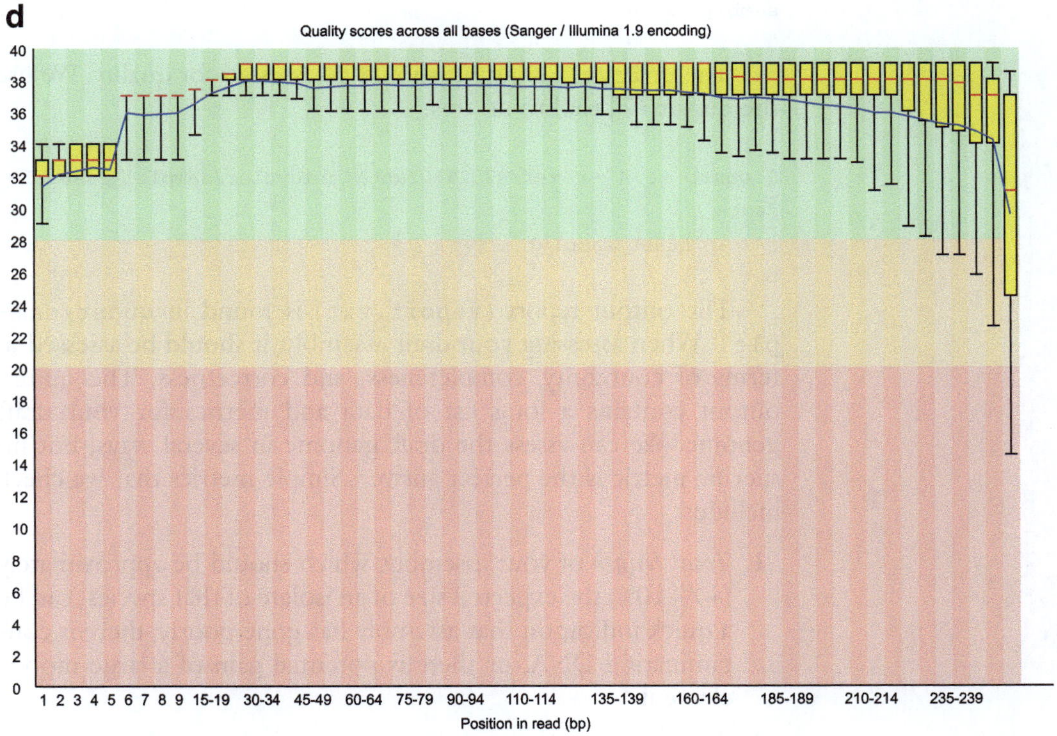

Fig. 1 (continued)

3.3 Create Genomes, Identify Genes, and Investigate Typing: Unicycler, Prokka, MLST

De novo assemblers are used to create large sequences (called contigs) by overlapping the trimmed reads and extracting the most likely genome sequence prior to fragmentation.

The `unicycler` [6] bioinformatic tool is designed specifically for bacterial samples. It can mix short (Illumina) and long reads (Nanopore and PacBio) and has inbuilt understanding of circular genomes. It is a composite of `unicycler` specific code and other tools, such as the `SPAdes` assembler.

It has a simple command line, using just the paired trimmed output from `fastp` (see above). It assembles the reads into contigs of minimum length 500bp (`--min_fasta_length`) and writes the files to the output folder (`-o`).

```
$ unicycler -t 4 --min_fasta_length 500 \
     -1 trim/sample1.p1.fastq.gz -2 trim/sample1.p2.fastq.gz \
      -o unicycler/sample1
```

The assembly contigs file (`assembly.fasta`) will be used in downstream analysis, and the corresponding output filename is derived from the input .fasta filename. It is therefore recommended to copy (or rename) the assembly contigs file to a more informative name (e.g., `sample1_assembly.fasta`).

```
$ cp unicycler/sample1/assembly.fasta unicycler/sample1_as-
sembly.fasta
```

The `unicycler` output should be assessed for quality. We use the `quast` [7] tool to do this:

```
$ quast -t 4 -r ref/R20291.fasta unicycler/sample1_assembly.
fasta \
    -o quast/sample1
```

The output report (`report.txt`) is found in `quast/sample1`. When assessing your draft assembly, it should be assessed in terms of contiguity, completeness, and correctness. The `quast` output contains a long list of data and metrics for your draft genome. We can assess the draft genome in several ways, but no specific metric is the perfect answer. Simple metrics that we check include:

1. *Total length* of your assembly, which should be approximately (+/−10%) the expected size of an isolate of that species, can be a quick indication that assembly has gone poorly, there is contaminating DNA, or there is potential gain of a large mobile element.
2. *Number of contigs*, with fewer contigs indicating better assembly. *C. difficile* has a single circular chromosome but can contain plasmids.

3. *Length of longest contig*, with the longest contig indicating a better assembly.
4. *N50*, which is a statistic calculated by ordering the contigs by size from largest to smallest and then identifying the length of the contig that straddles 50% of the genome length. Again, the longer, the better.

We have used a reference genome here, which `quast` compares your assembly to. Depending on how similar you expect your reference to be compared to your sample, comparisons to the reference can be useful:

1. *Genome fraction* (%), which measures the percentage of the reference genome covered by aligned assembly bases. A high value (e.g., >95%) indicates comprehensive coverage.
2. *Misassemblies* reflect the number of structural errors (inversions, relocations, translocations).
3. *Misassembled contigs* are contigs containing structural errors; lower values indicate better structural fidelity.
4. *NGA50* is a reference-aware version of N50. Calculated after splitting contigs at misassembly breakpoints while ensuring blocks align correctly to the reference. Higher NGA50 values reflect both continuity and structural accuracy.

To compare samples, it is useful order the contigs in a consistent manner. One method to do this is by matching contigs to a reference genome. The `abacas` [8] program does this using the `nucmer` tool [9] for the sequence alignment. The following command aligns our contigs created by `unicycler` to the R20291 *C. difficile* reference strain [10] downloaded above:

```
$ mkdir abacas
$ abacas.1.3.1.pl -r ref/R20291.fasta -p nucmer -c -d –N –a \
    -q unicycler/sample1_assembly.fa –o abacas/sample1
```

This command line tells `abacas` which reference genome to use (`-r`), which set of contigs to use (`-q`), and where to write the output fasta file (`-o`). The remaining options instruct `abacas` which matching software to use (`-p nucmer`) with default parameters (`-d`), which the genome is circular (`-c`), to not pad the output file with Ns (`-N`) and append any contigs that remained unmatched to the reference genome into the output file (`-a`).

The next step is to annotate the contigs with their likely genetic functional features, such as coding regions, genes, RNA, etc. The `prokka` [11] (short for prokaryotic annotation) program is ideal for this, as it combines BLAST [12] searches for common, conserved genes and hidden Markov models (via `hmmer` [13]) to identify more polymorphic genes. Running the following command will create a general feature format (GFF) annotation file composed of the identified genes and with the full sequence appended at the end:

```
$ prokka --cpus 4 --outdir prokka_sample1 --force --prefix sample1 --centre X \
      --compliant abacas/sample1.NoNs.fasta
```

The unpadded ordered contig file (`abacas/sample1.NoNs.fasta` created above) is processed into files, all beginning with sample (from the `--prefix` option) in the `prokka` directory (`--outdir` option). The `--force` option will allow `prokka` to overwrite any data in the output directory, as by default it will not destroy any previous output without this option. In addition to the GFF file, this directory contains the same information in multiple formats, including GenBank format (`prokka_sample1/sample1.gbk`) and a depreciated sequin format previously used for uploading to the GenBank database [14]. Also included are .txt, .log, and .err files that contain statistics on the analysis, a full log, and any errors encountered, respectively.

The annotated genome can be visualized using an online tool such as the Integrated Genome viewer (IGV, [15]). After navigating to the IGV page, click on the "Genome" menu button (top left), select "Local File...," and choose the `prokka/sample1.fna` fasta file. Next, click on "Track" menu button, select "Local File..." again, and choose the `prokka/sample1.gff` annotation file. The genome with genes annotated can be explored interactively.

Using mutation pattern in seven highly conserved genes, the multilocus sequence type (MLST) of a sample can be identified and placed in a wider *C. difficile* clade or lineage, with each clade exhibiting similar genetic and phenotypic profiles [16]. The alleles for these seven genes (*adk*, *atpA*, *dxr*, *glyA*, *recA*, *sodA*, and *tpi*) can be extracted from the contigs generated by `unicycler` or `abacas`. Missing or incomplete genes in the contigs can compromise the identification.

The `mlst` [17] command line tool automates searching for genes in contigs, identifying alleles, and assigning an MLST. It does this for over 100 typeable species (see `mlst --list`). There are two MLST databases for *C. difficile*; the first is the pubMLST Scheme [18], which is used throughout these analyses. It can be run on a single sample (1) or multiple samples at once by listing the fasta files and/or using wildcards (the `abacas/sample*.fasta` wildcard pattern matches all files that start with `sample` and end in `.fasta` within the `abacas` directory):

1. Single sample

```
$ mlst --legacy --nopath --threads 4 --scheme cdifficile abacas/sample1.fasta \
      > mlst/sample1.tsv
```

Table 1
MLST output. Multilocus sequence typing (MLST) of a sample genome file (sample1.fasta) shows the seven allele numbers in which unique combination results in a ST (sequence type)

File	Scheme	ST	*adk*	*atpA*	*dxr*	*glyA*	*recA*	*sodA*	*tpi*
sample1.fasta	cdifficile	1	1	1	1	10	1	3	5

2. Multiple samples

```
$ mlst --legacy --nopath --threads 4 --scheme cdifficile
ref/R20291.fasta \
     abacas/sample*.fasta > mlst/mlst_all.tsv
```

This command line runs the `mlst` tool giving the genic allele and MLST information (found in Table 1) on sample1. The output will be in the mlst folder and the .tsv files can be viewed, for example, by:

```
$ more mlst/sample1.tsv
```

or opening in Microsoft Excel or similar program.

Outbreak Analysis Have a look at your output for all the isolates. What does it tell you about these isolates? What can we infer from the MLST data? Do you think all the isolates are from a single outbreak? Are all isolates identical clones?

3.4 Multi-sample and Outbreak Analysis: Snippy, FastTree, ABRicate, Roary, chewBBACA

The next phase involves further analysis of multiple samples to assess their genetic relatedness and accessory genome content, supporting the investigation of potential outbreak events. For this section, we assume that every downloaded sample has been processed as per Subheadings 3.1 and 3.2, with the assembled genomes in the `unicycler` folder and the annotated genomes in the `prokka` folder.

`Snippy` [19] is a tool designed to identify single-nucleotide polymorphisms (SNPs) by mapping sequencing reads against a reference genome. For multisample analyses, the snippy-multi script, included with `Snippy`, simplifies the process. The script requires a tab-separated input file (`snippy_input.tab`), containing sample IDs and the corresponding post-`fastp` fastq filenames, formatted as follows:

```
sample1  trim/sample1.p1.fastq.gz  trim/sample1.p2.fastq.gz
sample2  trim/sample2.p1.fastq.gz  trim/sample2.p2.fastq.gz
```

To run snippy-multi:

```
$ snippy-multi --cpu 4 snippy_input.tab --ref ref/R20291.embl
> runme.sh
```

The `snippy-multi` command generates a script (`runme.sh`) that automates per-sample SNP analysis and core genome alignment using `snippy-core`. The ">" symbol instructs the computer to redirect any output from the program to the file (`runme.sh`) instead of writing it to the screen. Then we run this new script:

```
$ sh ./runme.sh
```

This generates per-sample SNP outputs stored in separate folders named after the sample ID provided in the tab input file. Additionally, core alignment files are produced, including `core.txt` (alignment or core-size statistics) and `core.full.aln` (whole genome SNP alignment).

To infer the phylogenetic relationships among the samples based on the core SNP alignment, the `core.full.aln` must first be cleaned to remove unwanted characters, replacing them with "N" to avoid artifacts in the downstream analysis, using the following commands:

```
$ snippy-clean_full_aln core.full.aln > clean.full.aln
$ run_gubbins.py -c 4 -p gubbins clean.full.aln
$ snp-sites -c gubbins.filtered_polymorphic_sites.fasta >
clean.core.aln
```

Following this, a phylogenetic tree is constructed using `FastTree` [20], which is a widely used tool for generating approximately-maximum-likelihood phylogenetic trees. The cleaning and tree-building can be done as follows:

```
$ FastTree -gtr -nt clean.core.aln > clean.core.tree
```

Tree visualization can be performed using a range of tools depending on the intended application. For interactive exploration, `GrapeTree` [21] provides a web-based platform that allows dynamic visualization and manipulation of phylogenetic trees. To use `GrapeTree` [22], click on "Load Files" button on the side bar and then click "Load Files" again in the pop-up window to upload the tree file (clean.core.tree). Metadata can be added by pasting into the text box in the pop-up window (*see* Fig. 2 for an example output).

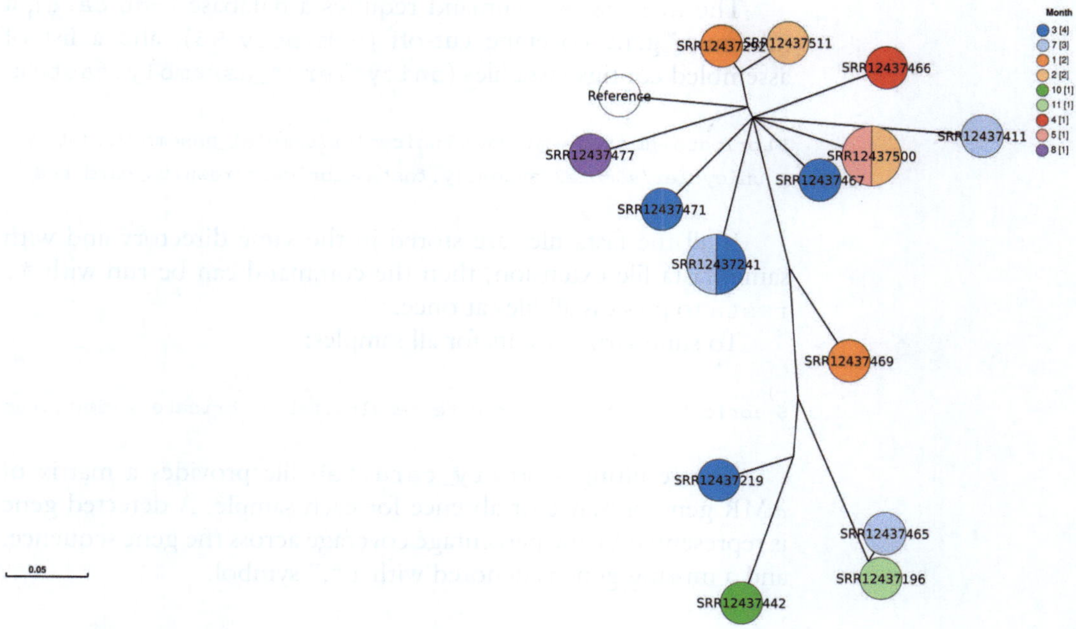

Fig. 2 Linking isolates by SNP distance. Phylogram [visualized in GrapeTree] using single-nucleotide point (SNP) mutations [snippy] distance matrix [fasttree] to link isolates. Isolates (circles) are linked by lines (distance, i.e., number of SNPs) and colored by metadata (month). Identical isolates result in larger divided circle

Outbreak Analysis The metadata contains information about the isolates including patient identifier, ward, month of isolation, and antimicrobial resistance (given as minimum inhibitory concentration data in μg/mL following EUCAST protocols). EUCAST MIC breakpoints are vancomycin S ≤ 2 μg/mL, R ≥ 2 2 μg/mL, and metronidazole S ≤ 2 μg/mL, R ≥ 2 2 μg/mL. What do the phylogenies tell you about the isolates? Do the isolates all look to be part of a single or multiple outbreak? When you look at the metadata for the isolates, what can you deduce from the phylogenies?

Exploration on antimicrobial resistance (AMR) profile may also provide insights into the outbreak events. `ABRicate` [23] performs mass screening of assembled contigs for AMR or virulence genes against different databases, including NCBI and CARD [24]. We can view a list with this command:

```
$ abricate --list
```

Database updates can be performed periodically, for example, to update the CARD database you would use:

```
$ abricate-get_db --db card
```

The abricate command requires a database (-db card), a minimum gene coverage cut-off (--mincov 95), and a list of assembled contigs fasta files (unicycler/*_assembly.fasta):

```
$ abricate -db card --mincov 95 unicycler/sample1_assembly.fasta \
  unicycler/sample2_assembly.fasta > abricate/results_card.tab
```

If all the fasta files are stored in the same directory and with same .fasta file extension, then the command can be run with *.fasta to process all files at once.

To summarize results for all samples:

```
$ abricate --summary abricate_results.tab > abricate_summary.tab
```

The resulting summary_card.tab file provides a matrix of AMR gene presence or absence for each sample. A detected gene is represented by the percentage coverage across the gene sequence, and a missing gene is denoted with a "." symbol.

Outbreak Analysis How does the abricate output compare to the clinical data present in the metadata.xls file? Do they match? Can you use the antimicrobial resistance genes (ARGs) to predict diagnostic laboratory data?

To complement the resistance gene analysis, pan-genome analysis using Roary [25] can be performed to identify core and accessory genes through the construction of a pan-genome. This analysis facilitates the examination of the gene gain and loss which may be related to the outbreak emergence. Roary requires annotated assemblies (GFF) produced by prokka as input. Multiple GFF files can be processed at once:

```
$ roary –e --mafft –p 4 -f roary -v prokka/*.gff
```

This command instructs roary to process all the GFF files in the prokka folder. Outputs are stored in the directory specified by the –f parameter. Summary statistics are reported in the summary_statistics.txt file. Roary also produces a gene_presence_absence.csv that contains detailed information, such as gene annotation and presence, for further manual exploration. Roary performs a quick alignment (-e —mafft) and generates the core gene alignment file (core_gene_alignment.aln). This .aln file can be used to build a phylogenetic tree using FastTree in a similar way as mentioned in the Snippy section.

ChewBBACA [26] is a set of bioinformatic tools for creating and evaluating core genome MLST (cgMLST) schemas that support molecular surveillance and outbreak investigations. These tools include schema creation and adaptation, allelic profiling, and cgMLST determination.

This tool needs extra files downloaded and processed once: the *C. difficile* cgMLST reference, which is a set of .fasta files:

```
$ wget "https://www.cgmlst.org/ncs/schema/Cdifficile1760/al-
leles/" \
      -O ref/cdiff_cgmlst.zip
$ unzip ref/cdiff_cgmlst.zip" -d ref/cdiff_cgmlst
```

Then, these data need to be processed to be ready to use in by the `ChewBACCA` tool with this command:

```
$ chewBBACA.py PrepExternalSchema --cpu 4 -i ref/cdiff_cgmlst
-o ref/cb_schema
```

The `PrepExternalSchema` command parameter prepares the coding sequence FASTA in the folder specified by the `-i` parameter for use and storage in the directory specified at the `-o` parameter. Alternatively, you can build a new schema from whole genome assemblies using the `CreateSchema` command parameter.

The pipeline in this chapter focuses on the allelic profiling functionality to analyze the outbreak events among the samples.

To streamline the allele calling process, all contig files (.fna), generated from the `prokka` annotation step, should be gathered into a directory, for example, a folder named `fna`:

```
$ cp prokka/*fna fna
```

Once organized, the allele calling can be performed using the AlleleCall function:

```
$ chewBBACA.py AlleleCall --cpu 4 -i fna -g ref/cb_schema -o
chewbbaca
```

This command instructs `chewBBACA` to scan each assembly contig file (`*.fna`) within the specified input directory (`-i fna`) against the prepared cgMLST schema (`-g ref/cb_schema`). The output files, including `results_statistics.tsv` (counts of allele classification type) and `results_alleles.tsv` (allelic profiles), are stored in the specified output directory (`-o chewbbaca`).

In the `results_alleles.tsv`, each row corresponds to a sample and column represents a locus defined in the cgMLST schema. The entries are allele identifiers, typically integers, that indicate which variant of each gene was found. When an allele cannot be confidently assigned, the entry is marked with a classification type to indicate the reason. These include:

Fig. 3 Linking isolates by cgMLST distance. Phylogram [visualized in Phyloviz] using core genome multilocus sequence typing (cgMLST) [chewBBACA] distance matrix [Phyloviz] to link isolates. Isolates (circles) are linked by lines (distance, i.e., number of shared alleles) and colored by metadata (month)

- *INF* (INFerred new alleles): A highly similar match is found, but no exact allele assignment can be made, often due to frameshifts or premature stop codons.
- *LNF* (Locus Not Found): No BLAST hits or not passing base score recalibration (BSR) threshold for the locus.
- *NIPH* (Non-Informative Paralogous Hit): Multiple coding DNA sequences (CDSs) from the same genome match the same schema locus, suggesting presence of paralogs.
- *PLOT3/PLOT5* (Possible Locus On the Tip): Indicates that there is an unaligned portion close to contigs 5′- or 3′- end. This could be due to genome fragmentation.

The `results_alleles.tsv` can be used to generate a phylogeny using PHYLOViZ [27] using "profile data" format with the phyloviz_meta.txt file. Note: The dataset has to be given a name to allow "Launch tree" (Fig. 3 shows an example output).

Outbreak Analysis What does the cgMLST profiles (results_alleles.tsv) tell you about the isolates? Is there a single cgMLST? If MLST profiles were identical, are the cgMLST? The cgMLST has

2147 alleles, how many differences between isolates would mean they are unrelated? What does the cgMLST phylogeny and metadata tell you, and how does this compare to the snippy phylogeny?

4 Conclusion

In the chapter, we have processed raw data that comes from sequencing machines to generate draft genomes for each sample. We have then contextualized these genomes individually via external data sources to annotate the genes, identify MLST STs, and classify AMR genes. Finally, we have assessed all the samples together to elucidate core/common genes, relatedness through phylogeny, and these have enabled us to answer questions about any outbreaks.

Every analysis is different, answering unique biological or epidemiological questions, with varying data quality/provenance, and requiring alternate tooling to process it, but this chapter outlines a strong foundational pipeline from data to answer that you can build upon as you analyze your own data.

Outbreak Analysis In this chapter we used WGS data from isolates that were identical by PCR-ribotyping (RT-027). When analyzed by MLST, these were all the same ST (ST1), but analysis of the AST data, such as the range of MIC values, suggests that these isolates are not identical. The snippy phylogeny quickly showed that there were differences between isolates, with a cluster of similar isolates and a long branch with slightly less similar isolates. Using the metadata you may have noticed:

- These two groups were separated by the ICU from which the patients were in when the *C. difficile* was isolated.
- The ICU1 isolates are very similar to the reference and each other so may represent an outbreak.
- P25 was sampled three times, with increasing MIC for some antibiotics. Further investigation may be able to link this to treatment, potentially.
- P07 and P12 had identical isolates taken over time, although P12 developed resistance to vancomycin, which may be related to changes in treatment, but that data is unavailable.
- P05 appeared to develop high levels of AMR.
- AMR genes were scarce and did not predict/match the antibiogram.
- The cgMLST profiles demonstrated few differences between isolates, indicating highly related isolates.

- The chewBBACA information could be used to identify shared genes, which could be used to develop a clone-specific PCR or the accessory genes could be used to look at the gain of specific genes.

References

1. Gargis AS et al (2023) Reference susceptibility testing and genomic surveillance of Clostridioides difficile, United States, 2012–17. Clin Infect Dis. https://doi.org/10.1093/cid/ciac817
2. Window Subsytem for Linux install: https://learn.microsoft.com/en-us/windows/wsl/install
3. Conda bioinformatics package management system: https://docs.anaconda.com/
4. Quality reporting package fastqc: https://www.bioinformatics.babraham.ac.uk/projects/fastqc/
5. Chen S et al (2018) fastp: An ultra-fast all-in-one FASTQ preprocessor. Bioinformatics. https://doi.org/10.1093/bioinformatics/bty560
6. Wick RR et al (2017) Unicycler: resolving bacterial genome assemblies from short and long sequencing reads. PLoS Comput Biol. https://doi.org/10.1371/journal.pcbi.1005595
7. Gurevich A et al (2013) QUAST: quality assessment tool for genome assemblies. Bioinformatics. https://doi.org/10.1093/bioinformatics/btt086
8. Assefa S et al (2009) ABACAS: algorithm-based automatic contiguation of assembled sequences. Bioinformatics. https://doi.org/10.1093/bioinformatics/btp347
9. Guillaume M et al (2018) MUMmer4: a fast and versatile genome alignment system. PLoS Comput Biol. https://doi.org/10.1371/journal.pcbi.1005944
10. R20291 reference strain: https://www.ebi.ac.uk/ena/browser/api/fasta/FN545816.1
11. Seemann T (2014) Prokka: rapid prokaryotic genome annotation. Bioinformatics. https://doi.org/10.1093/bioinformatics/btu153
12. Camacho C et al (2009) BLAST+: architecture and applications. BMC Bioinformatics. https://doi.org/10.1186/1471-2105-10-421
13. Eddy SR (1998) Profile hidden Markov models. Bioinformatics. https://doi.org/10.1093/bioinformatics/14.9.755
14. Benson DA et al (2013) GenBank. Nucleic Acids Res. https://doi.org/10.1093/nar/gks1195
15. Integrate Genome Viewer software: https://igv.org/app
16. Stabler RA et al (2012) Macro and micro diversity of Clostridium difficile isolates from diverse sources and geographical locations. PLoS One. https://doi.org/10.1371/journal.pone.0031559
17. Sleemann T. mlst. Guthub https://github.com/tseemann/mlst
18. http://pubmlst.org/cdifficile/ fromJolley KA, Maiden MC (2010) BIGSdb: scalable analysis of bacterial genome variation at the population level. BMC Bioinformatics. https://doi.org/10.1186/1471-2105-11-595
19. Sleemann T, Snippy. Github. https://github.com/tseemann/snippy
20. FastTree: Price MN et al (2010) FastTree 2 – approximately maximum-likelihood trees for large alignments. PLoS One. https://doi.org/10.1371/journal.pone.0009490
21. GrapeTree: Zhou Z et al (2018) GrapeTree: visualization of core genomic relationships among 100,000 bacterial pathogens. Genome Res. https://doi.org/10.1101/gr.232397.117
22. Phylogeny plotting tool GrapeTree: https://achtman-lab.github.io/GrapeTree/MSTree_holder.html
23. Seemann T, Abricate, Github. https://github.com/tseemann/abricate
24. Jia B et al (2017) CARD 2017: expansion and model-centric curation of the comprehensive antibiotic resistance database. Nucleic Acids Res. https://doi.org/10.1093/nar/gkw1004
25. Page AJ et al (2017) Roary: rapid large-scale prokaryote pan genome analysis. Bioinformatics. https://doi.org/10.1093/bioinformatics/btv421
26. Silva M et al (2018) chewBBACA: a complete suite for gene-by-gene schema creation and strain identification. Microbial Genomics. https://doi.org/10.1099/mgen.0.000166
27. Web-based phylogeny visualistion tool: https://online.phyloviz.net/index

Chapter 2

Generating Targeted Mutants Using a Riboswitch to Control Plasmid Behavior During Allelic Exchange Mutagenesis

Joshua N. Brehm and Joseph A. Sorg

Abstract

Efficient and markerless allele-coupled exchange (ACE) mutagenesis is possible in a wild-type *C. difficile* background by using a synthetic riboswitch to indirectly control replication of the mutagenesis plasmid. Controlling plasmid replication gives the user the ability to introduce selective pressure for recombination of the mutagenesis plasmid into the chromosome (integration) followed by recombination of plasmid DNA out of the chromosome (excision) reliably. ACE has been successfully used in *C. difficile* research for over a decade, mostly with unstably replicating plasmids using toxic pyrimidine precursors or toxin/antitoxin systems to select for recombination events. The advantages of using a riboswitch to control plasmid replication include the ability to begin with a wild-type parent strain, use of a higher copy number origin of replication, and generation of mutants on rich medium. A wild-type progenitor means there is no need to complement *pyrE* into newly made mutants, increasing ease of use. The inclusion of the more stable, pCD6-derived origin of replication in the mutagenesis vector should increase efficiency of conjugal transfer, and its higher plasmid copy number should increase the frequency of crossover events. The efficacy of the riboswitch in rich media also avoids the use of minimal or defined media, increasing growth rates and ease of use.

Key words Allelic exchange mutagenesis, Riboswitch, Targeted mutations

1 Introduction

The ability to efficiently generate markerless, targeted mutations in a bacterium is an essential tool for many experimental designs. ACE [1–3] and CRISPR-Cas [4–7] mutagenesis systems have been successfully used to generate novel strains of *C. difficile*, but mutagenesis remains a bottleneck in many experimental processes. ACE systems suffer from difficulties in screening for rare crossover events, low conjugation efficiencies (due to encoded toxins or unstable origins), and the required use of defined minimal media. CRISPR mutagenesis systems are simpler to use, requiring only a single screen for mutants, but can also suffer from low conjugation efficiencies (due to plasmid size and toxicity) and require the

Kurt H. Piepenbrink and Jennifer M. Auchtung (eds.), *Clostridioides difficile: Methods and Protocols*, Methods in Molecular Biology, vol. 3046, https://doi.org/10.1007/978-1-0716-5328-9_2,

identification of a functional guide RNA target within the targeted region (many theoretically targeted gRNA sequences fail to create mutants). Our recently introduced ACE system alleviates many of these issues. Cells with chromosomally integrated plasmid DNA (integrants) are easier to identify due to higher selective pressure and faster growth, resulting in a higher frequency of colonies which are noticeably and reproducibly larger. Recombinants which have lost plasmid DNA from the chromosome are quicker to identify due to selection occurring in rich medium without the addition of toxins or toxin-inducing compounds. The system also avoids unnecessary rounds of cloning associated with the process of identifying functional gRNA for CRISPR-based systems as well as the decreased rates of conjugal plasmid transfer seen with larger plasmids.

A riboswitch is a sequence which, when transcribed, can form a secondary structure capable of regulating RNA or protein synthesis. The riboswitch used here [8] forms a stem loop that blocks the Shine–Dalgarno sequence, thereby preventing assembly of the ribosome. This blockage can be alleviated by a conformational change resulting from the binding of theophylline to the riboswitch, thus allowing translation to occur normally. Below, the described mutagenesis system acts through the toggling of this riboswitch cloned into the described plasmid directly upstream of *repA*. RepA is necessary for initiating plasmid replication. As a result of the riboswitch, translation of RepA (and, thus, replication of the plasmid) can be controlled by supplementation of theophylline in the growth medium.

By restricting plasmid replication while maintaining selective pressure with antibiotics, cells which integrate the plasmid into their chromosome can continue growth in the presence of antibiotics despite the depletion of the reservoir of free plasmid copies (due to the presence of the antibiotic resistance marker encoded on the plasmid, now integrated on the chromosome). This results in efficient and reliable identification of integration events on rich medium in a wild-type background (integrant colonies grow normally while non-crossovers form pinpoint colonies). Next, integrants are repassaged and isolated on antibiotic media lacking theophylline (to ensure that there are few free plasmid copies). Cells in which plasmid DNA has excised from the chromosome are then identified via passaging of integrant cells on media containing theophylline. We hypothesize that the presence of a second, highly active, origin of replication interferes with normal chromosomal replication, providing a selective pressure for removal of plasmid DNA from the chromosome. The result is a clear differentiation between diffuse integrant growth and more robust growth from cells which have excised the plasmid DNA. The identified colonies can then be tested for the desired mutation.

2 Materials

2.1 Media

Required media listed here, instructions included below:

- BHIS agar.
- BHIS–thiamphenicol (Tm) agar.
- BHIS–theophylline (Theo) agar.
- BHIS–Tm + Theo + kanamycin (Kan) + D-cycloserine agar (TTKC).
- LB–chloramphenicol (Cm) + ampicillin (Amp) agar.
- LB–Cm agar.
- LB–Cm + 1% sucrose agar.
- LB–Cm Amp broth.

BHIS agar—Dissolve the following in 800 mL dH_2O: 37 g brain–heart infusion and 5 g yeast extract. Bring total volume up to 1 L with additional dH_2O, and add 15 g agar. Autoclave, and allow to cool slightly. While medium is cooling, dissolve 1 g L-cysteine in 10 mL dH_2O and filter-sterilize. Add the L-cysteine solution to the medium before pouring.

LB agar—Dissolve 25 g LB powder in 800 mL dH_2O, and bring up to 1 L with dH_2O. Add 15 g agar and autoclave. Add supplements while cooling, just before pouring.

LB broth—Dissolve 25 g LB powder in 800 mL dH_2O, bring up to 1 L with dH_2O, and autoclave.

Supplements

- Theophylline—2 g/L added as powder to warm medium after autoclaving (due to low solubility and difficulties in making a concentrated stock).
- Sucrose—Filter-sterilized solution added to warm medium after autoclaving to final concentration of 1%.
- Antibiotics—Added after autoclaving to respective final concentrations:
- Thiamphenicol—5 μg/mL.
- Kanamycin—50 μg/mL.
- D-cycloserine—250 μg/mL.
- Chloramphenicol—20 μg/mL.
- Ampicillin—100 μg/mL.

2.2 Bacterial Strains and Plasmids

- *E. coli* DH5α for transformations and cloning.
- *E. coli* DH5α pJB94 for plasmid storage/purification:

  - Available through AddGene (plasmid #208752).
- *E. coli* HB101 pRK24, or similar strain (e.g., *E.* coli CA434 or *E. coli* sExpress) [9], for conjugal transfer.

2.3 Molecular Biology Reagents/Enzymes

Cloning Reagents

1. High-fidelity DNA polymerase and required reagents:
   - Phusion, Phusion Plus, Q5, or similar
2. Cloning oligonucleotides—Order oligos to PCR amplify regions 500 base pairs upstream and 500 base pairs downstream of your desired mutation. They must have suitable 5′ ends for your preferred plasmid assembly method (e.g., Gibson Assembly, ligation). If your desired mutation includes an insertion, you will need oligos to amplify that as well.
3. Chromosomal test oligonucelotide—It is necessary to use an oligo which binds close to the locus of the desired mutation but which will not bind to the regions of homology on your finished mutagenesis plasmid. This allows PCR testing for plasmid integration and chromosomal mutations without undesired amplification of plasmid DNA. This is normally designed to bind directly upstream of the upstream homology fragment 5′ primer.
4. M13F/M13R standard oligonucleotides—Common primers, used here to bind to plasmid DNA when testing for integration events or free plasmid.
5. NotI and XhoI restriction enzymes and corresponding buffers.

3 Methods

3.1 Cloning

1. Digest pJB94 with NotI and XhoI and purify. This will remove the ~2.5-kb *sacB* gene and allow for the insertion of the necessary regions of homology/insert.
2. Amplify and purify the relevant PCR fragments for your desired plasmid assembly method (e.g., Gibson assembly, ligation) and assemble your plasmid.
3. Transform the plasmid into *E. coli* DH5α and spread on LB–Cm agar medium supplemented with 1% sucrose. Sucrose is a selective agent which kills cells that have taken up undigested plasmid (sucrose is toxic to cells containing *sacB*) from the sample generated in step 1.1. It reduces the number of false positives in the transformation.
4. Streak several transformant colonies to the LB–Cm agar medium and incubate overnight. Purify and test for correct

plasmid assembly via PCR or band size after digestion with NotI/XhoI. Sequence plasmid DNA and freeze at least one confirmed isolate for storage.

3.2 Conjugal Transfer

1. Transform the purified plasmid into *E. coli* HB101 pRK24 and spread on LB–Cm Amp agar (we have success with *E. coli* HB101 pRK24, but pJB94 also contains the Tn*916* recognition sequence commonly used for *B. subtilis* transfer). *E. coli* CA434 can also be used.
2. Simultaneously culture *E. coli* HB101 pRK24 + your plasmid (aerobically in LB–Cm Amp liquid medium) and the *C. difficile* parent strain (anaerobically in BHIS liquid medium) to an OD_{600} of ~1.
3. Spin 1 mL of the *E. coli* culture in a 1.5-mL tube at 5000× *g* for 30 s. Remove supernatant and pass the cell pellet into the anaerobic chamber.
4. Gently resuspend the *E. coli* pellet in 200 μL of the *C. difficile* culture, and then drop the entire mixture in ~20 μL spots across a plate of BHIS–Theo agar medium. Let the spots dry, then invert the plate, and incubate overnight (at least 16 h).
5. Scrape growth from the spotted plate into 1-mL BHIS liquid medium, then pipette, and spread mixture evenly over 5 plates of BHIS–TTKC agar. Incubate overnight and check for transconjugant colonies daily for 72 h. Streak potential transconjugants to reduced BHIS–TTKC agar and incubate overnight.
6. Use PCR to check isolates for the presence of the plasmid using the M13F/M13R oligonucleotides as well as amplifying a chromosomal marker (e.g., *tcdB*) to ensure growth is *C. difficile*.
7. Freeze growth from at least one confirmed transconjugant for potential later use.

3.3 Integrant Isolation

1. Streak a single transconjugant colony onto reduced BHIS–Tm agar and incubate overnight (*see* **Notes 1–3**).
2. Closely investigate the agar to identify colonies which stand out from the diffuse background growth. These might be fully isolated colonies which grow to a larger size, or larger colonies surrounded by, and overlapping with, poor growth. Normally, these colonies only become visible after overnight incubation, but tend to become far more noticeable after 48 h of growth. Under these conditions, the background growth is unable to efficiently pass the antibiotic resistance marker coded by the plasmid to daughter cells (lack of theophylline in the medium restricts plasmid replication). However, the larger colonies with the plasmid integrated into the chromosome continue to grow at a more normal rate due to replication of *catP* with the chromosome (*see* **Note 4**).

3. Streak potential integrant colonies onto reduced BHIS–Tm agar and incubate overnight. After overnight incubation, colonies should appear similar to what is observed for healthy *C. difficile* growth on rich medium. If growth still appears diffuse and generally unhealthy, the chosen colonies were not integrants, and **step 1** should be repeated.
4. Streak an isolated integrant colony onto reduced BHIS–Tm agar and incubate overnight. This last passage on BHIS–Tm is to ensure that free plasmid is lost from cells before continuing the protocol. Although the theophylline riboswitch is efficient at blocking translation of *repA*, the unknown stability of pre-existing RepA as well as the presence of multiple plasmid copies per cell mean that free plasmid is still present in the population during earlier passages.
5. Cells in which plasmid integration has occurred display a colony morphology that can be consistently identified visually with practice. However, when learning the protocol or if issues are encountered, it is always advisable to test cells to confirm integration events. This is done by testing candidate strains via PCR with the chromosomal test primer and M13F. Two different band sizes are possible, depending on whether your crossover event happened in the upstream or downstream homology region, but the presence of either band signifies a plasmid integration (*see* **Note 5**).

3.4 Double Crossover Isolation

1. Streak an isolated integrant colony onto reduced BHIS–Theo agar and incubate overnight. The reintroduction of theophylline to the growth medium allows translation of *repA* and causes the plasmid origin to attempt a return to multiple copies. Because this origin of replication is integrated into the chromosome, the end result is selective pressure for excision of the plasmid DNA from the chromosome (either regenerating the wild-type sequence or the replacement of the chromosomal allele with what was on the plasmid).
2. Growth should be sparse due to the strong selection, with isolated colonies likely in the second quadrant. Those colonies which can grow should have undergone excision. Pick and streak several of these to BHIS–Theo and incubate overnight.

3.5 Mutant Confirmation

1. Perform colony PCR on your isolated candidates. Use the downstream homology 3′ oligo from cloning paired with your 5′ test oligo which does not bind to plasmid DNA. This primer pair will result in amplification from both wild-type and mutant DNA, with a size difference depending on your introduced mutation. When testing for a deletion, the band size is ~1000 base pairs for a mutant and ~1000 base pairs + the length of the targeted gene for wild type.

2. Once a mutant is identified, it is important to confirm the absence of plasmid DNA in the cell. This can be done by streaking an isolated colony to BHIS agar media with and without supplemented Tm in parallel, or by testing for the presence of *catP* via PCR along with your mutant PCR screen, or both (*see* **Note 5**).

3.6 Curing of Free Plasmid

1. If all strain isolates which contain the desired mutation also contain plasmid DNA, streak several of them for isolation on reduced BHIS–Theo agar medium and incubate overnight to ensure excision of plasmid DNA from cell chromosomes.
2. Streak an isolated colony to reduced BHIS agar medium and incubate overnight, and then repeat. These passages without theophylline should result in the loss of free plasmid from your mutant strain.
3. Streak several colonies to BHIS and BHIS–Tm agar media in parallel to test for plasmid presence. If no cured isolates are detected, we recommend starting again from Subheading 3.1 (*see* **Note 6**).

4 Notes

1. It is crucial to practice good isolation streak technique for all steps of this protocol. Colonies must be isolated to observe growth morphology changes. The more isolated colonies per sample, the higher the probability of detecting the desired phenotype.
2. It is advisable to perform multiple replicates of all stages of the mutagenesis protocol in parallel (i.e., streak 6 transconjugants onto 6 plates of BHIS Tm agar), especially when learning the steps for the first time. This gives the person performing the protocol a better understanding of the colony morphologies to look for and demonstrates the natural variability in rates of rare crossover events among biological and technical replicates.
3. We use 5 μg/mL thiamphenicol in Subheading 3 because the desired cell genotype (one copy of plasmid sourced DNA integrated into the chromosome with no free plasmid) only contains one copy of the resistance marker *catP*. In *E. coli*, chloramphenicol resistance has been shown to depend on copy number of the resistance marker. The reduced thiamphenicol concentration still selects against wild-type *C. difficile*, but should allow for healthier growth of integrant colonies.
4. If no integrant colonies are detected after 48 h of growth, re-streak a small loopful of diffuse growth onto new, reduced BHIS–Tm agar and wait another 24–48 h. If there are still no integrants detected, we recommend starting again from **step 1**.

5. If difficulties are encountered in removing free plasmid from a confirmed mutant isolate later in the protocol, it is advisable to test the integrant samples from Subheading 3.5 for free plasmid as well as integrant events. This can be done with primers M13F and M13R, which will only amplify successfully from a plasmid that has not recombined at either homology arm. These free plasmids will not change the efficacy of selection for excision, but can be a problem to remove later in the protocol and should be avoided.
6. Difficulty removing unwanted plasmid DNA in a confirmed mutant strain is a common issue. Although unpublished, we have had great success cloning *codA* into the backbone of the mutagenesis vector and supplementing 50 μg/mL 5-fluorocytosine (5-FC) into the BHIS Theo agar used in **Note 1**. The selection against *codA* is weak in BHIS but does result in a visually smaller colony compared to those which still contain plasmid DNA. The majority of colonies observed after overnight growth on BHIS Theo 5-FC (seen in **Note 2**) tend to have both excised the plasmid from the chromosome and lost it entirely. Minimal medium supplemented with theophylline and 5-FC will result in stronger selection against *codA* than in BHIS, but at the cost of longer growth times and difficulties in culturing some strains (e.g., metabolic mutants). This plasmid, pJB127, also is deposited in Addgene (Addgene ID: 248630).

References

1. Cartman ST et al (2012) Precise manipulation of the Clostridium difficile chromosome reveals a lack of association between the tcdC genotype and toxin production. Appl Environ Microbiol 78(13):4683–4690
2. Ng YK et al (2013) Expanding the repertoire of gene tools for precise manipulation of the Clostridium difficile genome: allelic exchange using pyrE alleles. PLoS One 8(2):e56051
3. Peltier J et al (2020) Type I toxin-antitoxin systems contribute to the maintenance of mobile genetic elements in Clostridioides difficile. Commun Biol 3(1):718
4. Brehm JN, Sorg JA (2022) Plasmid sequence and availability for an improved Clostridioides difficile CRISPR-Cas9 mutagenesis system. Microbiol Resour Announc 11(12):e0083322
5. Hong W et al (2018) Multiplexed CRISPR-Cpf1-mediated genome editing in Clostridium difficile toward the understanding of pathogenesis of C. difficile infection. ACS Synth Biol 7(6): 1588–1600
6. Maikova A et al (2019) Using an endogenous CRISPR-Cas system for genome editing in the human pathogen Clostridium difficile. Appl Environ Microbiol 85(20):e01416
7. McAllister KN et al (2017) Using CRISPR-Cas9-mediated genome editing to generate C. difficile mutants defective in selenoproteins synthesis. Sci Rep 7(1):14672
8. Suess B et al (2004) A theophylline responsive riboswitch based on helix slipping controls gene expression in vivo. Nucleic Acids Res 32(4): 1610–1614
9. Woods C et al (2019) A novel conjugal donor strain for improved DNA transfer into Clostridium spp. Anaerobe 59:184–191

Chapter 3

Use of the Golden Gate Cloning Technique for Plasmid Construction to Generate Conditional Lethal Mutants in *C. difficile*

Jeanne Malet-Villemagne, Marie-Emeline Zielinski, Assilina Parfut, and Thomas Candela

Abstract

Conditional lethal mutants are a valuable tool for studying essential genes in bacteria, as they allow for the controlled expression of the gene of interest under specific conditions. In this study, we present a protocol for generating conditional lethal mutants in *Clostridioides difficile* using the Golden Gate cloning technique. The strategy involves two steps: (1) introducing an inducible copy of the gene of interest (*lcpB,* in the example) into the erythromycin resistance locus of *C. difficile* by allelic exchange and (2) deleting the native copy of the gene(s) of interest (*lcpA* and *lcpB* in the example) by allelic exchange. The protocol utilizes pseudo-suicide vectors carrying antibiotic resistance genes and the Golden Gate assembly method for efficient plasmid construction. The expression of the gene of interest can be modulated by varying the concentration of the inducer anhydrotetracycline, allowing for the observation of phenotypes associated with low gene expression. This technique provides a powerful approach for understanding the functions of essential genes in *C. difficile* and can be adapted for studying essential genes in other bacteria.

Key words *Clostridioides difficile*, Essential gene, Conditional lethal mutant, Golden Gate

1 Introduction

Gene function is usually determined using the corresponding mutants. Therefore, it is not possible to identify the functions of essential genes and alternative methods have been proposed for this purpose. Essential genes cannot be studied by disruption, as in other genes, because by definition, their disruption leads to bacterial death. Essential genes are very complicated to study but have essential roles in bacterial cells. Understanding their roles is of great interest to understand the numerous mechanisms of bacterial life (multiplication, sporulation, metabolism, stress responses, antibiotic resistance, etc.). The most common strategy is to target essential gene expression using antisense RNA, or similar strategies

Kurt H. Piepenbrink and Jennifer M. Auchtung (eds.), *Clostridioides difficile: Methods and Protocols*, Methods in Molecular Biology, vol. 3046, https://doi.org/10.1007/978-1-0716-5328-9_3,

[1, 2]. However, these strategies are not fully effective because the expression of the targeted gene usually decreases, but is not abolished. In addition, the antisense strategy may modify the expression of other genes that are transcribed in the operon. The techniques based on CRISPRi [3, 4] have similar limitations. Another approach for studying essential genes involves creation of conditional lethal mutants. Conditional lethal mutants are those that are unable to grow under specific conditions determined by the experimenter, but can grow under permissive growth conditions. The construction of conditional lethal mutants of essential genes of interest is a way to study them by lowering their expression while keeping the mutant alive. These mutants allow researchers to control gene expression and observe the effects of the inactivation of essential genes under specific conditions. In addition, clean deletion of the targeted gene allows the expression of other genes present in potential operons. Therefore, the latter technique is more appropriate for studying essential genes, although it is difficult to establish. A conditional lethal mutant technique has recently been published for *Clostridioides difficile* [5].

C. difficile is a ubiquitous bacterium present in various environments (soil, water, vegetables, meat products, and seafood) that poses a problem for human and animal health. *C. difficile* is a major cause of antibiotic-associated diarrhea in developed countries (for reviews: [6–8]). Elucidation of the mechanisms of *C. difficile* infection and the discovery of new therapeutic strategies must involve the study of essential genes. For this purpose, new molecular biology tools are needed and developed recently. These new tools include inducible promoter systems, such as the P_{tet} promoter [9], toxin–antitoxin allelic exchange system [10], and the renewal of the Golden Gate technique (https://www.neb.com/en/golden-gate/golden-gate?srsltid=AfmBOoqiw32Z-WLjxnuOQo7qU8SwoUjs3reEBaCy8y7HVyyoiTcXuMgu).

In this protocol, we show the use of the conditional technique to delete the two *lcp* genes of *C. difficile* that are essential for *C. difficile* growth [5].

2 Materials

2.1 Strains and Culture Conditions

2.1.1 Strains Used in the Study

Two *E. coli* strains (Table 1) are required for this protocol: the TG1 strain is used for complementation and deletion plasmid production, and the HB101 pRK24 strain allows the transfer of shuttle plasmids to *C. difficile* 630 by conjugation. Both strains were cultivated in Luria–Bertani (LB) medium at 37 °C. This medium is routinely used for *E. coli* and is commercially available (BD Difco).

The native 630 strain (Table 1) of *C. difficile* [11] is used to generate conditional lethal mutants. This clinical strain is essential

Table 1
Strains of the study

Strain name	Genotype	Source or reference
Escherichia coli		
TG1	*E. coli* k12 (F′, *tra* D36, *lacIq*, Δ*lacZ*, MIS, *pro* A+B+/SupE, Δ(*hdsM-mcrB*))	Laboratory stock
HB101 pRK24	*E. coli* (pRK24) (F - Δ(*gpt-proA*) 62 Leu B6 *gln* V44 *ara*-14 *galK*2 *lac*Y1 Δ(*mcrC-mrr*) *rps* L20 (*srt*^{r}) *xyl*-5 *mlt*-1 *rec* A13, pRK24	Laboratory stock
Clostridioides difficile		
630	Clinical strain, Erm^R	[11]
JMV2 (630 P_{tet} - *lcpB*)	Derivative of 630 strain, Erm^S, Δ(*CD630_20100*, *CD630_20091*, *CD630_20090*, *CD630_20080*, *CD630_20071*, *CD630_20070*):: P_{tet}-*lcpB*	[5]
JMV6 (Δ*lcpA* Δ*lcpB* P_{tet}-*lcpB*)	JMV2 Δ*lcpA* Δ*lcpB*	[5]

for this technique because its *ermB* locus is intact (contrary to the widely used 630Δ*erm* strain); therefore, the strain is erythromycin-resistant. A supplementary inducible copy of the gene of interest will be inserted into this locus to facilitate screening (clones with inserted genes are erythromycin sensitive). Visual prescreening by colony restreaking on brain–heart infusion (BHI) plates with erythromycin was possible, followed by polymerase chain reaction (PCR) confirmation. *C. difficile* 630 strain should be grown in BHI medium at 37 °C under anaerobic conditions. This medium is routinely used for *C. difficile* and is commercially available (BD Difco). The BHI medium should be supplemented with 1% defibrinated horse blood or antibiotics at the following concentrations: thiamphenicol (Th; 7.5 μg mL^{-1}), aztreonam (Az; 16 μg mL^{-1}, used to kill parental *Escherichia coli* during the conjugation process), or erythromycin (Er; 5 μg mL^{-1}). Anhydrotetracycline (ATc) is used to induce the P_{tet} promoter (concentrations from 5 to 250 ng mL^{-1}). LB medium should be supplemented with ampicillin (Amp; 100 μg mL^{-1}), chloramphenicol (Cm; 25 μg mL^{-1}), kanamycin (Kn; 40 μg mL^{-1}), spectinomycin (Spec; 100 μg mL^{-1}), and erythromycin (Er; 150 μg mL^{-1}).

2.2 Plasmids and Primers

The plasmids and primers to be used are listed in Table 2.

2.3 Molecular Biology

We have had success with plasmid extraction, PCR purification, and gel extraction performed using the Omega E.Z.N.A plasmid DNA cycle pure kit and gel extraction kit, respectively, according to the

Table 2
Plasmids and primers used in this study. Primer sequences in italics correspond to nucleotides that do not match the target sequence. The BsaI enzyme site is underlined

Plasmids or primer name	Genotype/primer sequence	Source or reference
Plasmids		
pMSR	Circular cloning vector, 5624 nucleotides, *catP*, α*lacZ*, P_{tet}-*CD2517* (toxin), pseudo-suicide plasmid, Cm^R	Gift from J. Peltier [10]
pJV7	pMSRΩ*aadA*, spectinomycin resistance gene flanked by BsaI sites, Cm^R, and Sp^R	[5]
pMEZ2	pJV7 derivative, kanamycin resistance gene flanked by BsmBI sites, CmR, and KnR	This work
pJV8	pMSR derivative, plasmid used for *erm* locus deletion, spectinomycin resistance gene flanked by BsaI sites, Cm^R, and Sp^R	[5]
pMEZ9	pJV8 derivative, plasmid used for *erm* locus deletion, kanamycin resistance gene flanked by BsmBI sites, Cm^R, and Kn^R	This work
pJV10	pJV7Δ*catP*Ω*ermB*, Erm^R, and Sp^R	[5]
pMEZ3	pJV10 derivative, kanamycin resistance gene flanked by BsmBI sites, Erm^R, and Kn^R	This work
pJV13	pJV10 derivative, plasmid used for *lcp* region deletion, Erm^R	[5]
pJV27	pJV8ΩP_{tet}-*lcpB*, Cm^R	[5]
pJV17	pJV8ΔcatPΩermB, ErmR, and SpR	This work
pMEZ4	pJV17 derivative, kanamycin resistance gene flanked by BsmBI sites, Erm^R, and Kn^R	This work
Primers		
JV58	*CGATAG GGTCTC GTTGC*GTCACCAAATACCATAGTTTCTT	[5]
JV69	*AGTACC GGTCTC* CTGCCTTGATGGTATAACATCAACACC	
JV70	*TTCCTG GGTCTC CCC*TAATATCCCCTACTTTCTAAATTTTTTAAT	
JV71	*TTCCTG GGTCTC* CTAGGCCGGCCAAGTGGGCAA	
JV72	*TTCCTG GGTCTC CTTCT*TAGGGTAACAAAAACACCGTATTTC TACGATGT	
JV73	*TTCCTG GGTCTC* CAGAAAATTCAACATAAAGTTTATTAAAAAG TATAAGATTAATTACT	
JV85	GCACTTTTCATCATTTCCACATCATTTAAC	
JV87	CAAATTCAGATACAGTAGTATTAGTAAATG	
JV90	GTACAAGGTACACTTGCAAAGTAGTGGTC	
JV91	CAAGTTCATCACGCAGTATGTGACGG	
JV99	GGCATGGCACATCAGTAAAAATTGAATAC	
JV100	CGGTCACGGTGTAATCTTCTGTGACTGCC	
JV136	*GGCTACGGTCTCTTGCC*AAAAATAAGAAGCCTGCATTTGC	
JV137	*GGCTACGGTCTCT*TCCTTTACTGCAGGAGCTC	
JV138	*GGCTACGGTCTCTAGGAGAAAATT*TTTGTCAAAATTAAAGAAA TTTGTTATAC	
JV139	*GGCTACGGTCTCCTTGC*TTATTGTTTAAACTCTATGTCATTAAA TATAAAATC	
1998	*GAATTC* GAATATTACTACCAAGAAAGCCAGTAG	This work
1999	*GAATTC* GACATATTACACGATTTTATATTTAATGAC	This work

manufacturer's instructions. We have performed PCR for the gene amplification of genomic DNA and mutant screening of *C. difficile* using high-fidelity Phusion DNA polymerase (NEB Biolabs). *Taq* DNA polymerase can be used to screen *E. coli* clones (2x rapid Taq Master Mix, Vazyme). All DNA samples can be electrophoresed for verification on standard agarose gels at 1.0% (w/v). We have used a T100 thermocycler for the PCRs. We perform sequencing using Eurofins Genomics.

3 Methods

In the last decade, the genetic manipulation of *C. difficile* has been extensively improved, allowing the understanding of numerous processes and mechanisms. Different gene disruption techniques are now routinely used (group II introns, such as ClosTron, site-directed mutagenesis, mutation by homologous recombination, Crispr-Cas [10] methods, for review [12]), but none of them allow the proper study of essential genes to explore their functions. We developed a conditional lethal mutation technique that allows the study of essential genes in *C. difficile*. The strategy consists of two steps, as shown in Fig. 1: (**step 1**) introduction of an inducible copy of the gene of interest (in our example, *lcpB*) in the erythromycin resistance locus of *C. difficile* by allelic exchange and (**step 2**) deletion of the native copy of the gene of interest (in our example, both *lcpA* and *lcpB* [5, 13]) by allelic exchange.

3.1 Construction of the Plasmid to Perform Step 1

Allelic exchange is a reliable mutation technique that is frequently used in *C. difficile*. Due to the low frequency of gene transfer in *C. difficile*, the use of suicide plasmids (which are unable to replicate in *C. difficile*) is not possible. Therefore, use of "pseudo-suicide" vectors is recommended. Such vectors carry a gene that confers resistance to a specific antibiotic (e.g., *catP* encoding resistance to thiamphenicol), and their replication in the presence of the antibiotic is defective enough to be a significant growth disadvantage. We constructed a new plasmid suitable for the Golden Gate technique and derived it from pMSR [10], which contains the upstream region of *the ermB* locus, a spectinomycin cassette flanked by BsaI restriction sites, and the downstream sequence of *the ermB* locus, named pJV8 [5] (*see* **Note 1**) (Fig. 2). This plasmid allowed for integration of any chosen sequence at the *ermB* locus. For the mutation of *lcp* genes, we cloned the P_{tet} promoter and *lcpB* gene into the pJV8 plasmid. The P_{tet} promoter was amplified from 10 ng of pRPF1858 using JV136/JV137 primers (10 μM) and Phusion polymerase in GC buffer. *lcpB* was amplified from 10 ng of genomic DNA of 630 strain using JV138/JV139 primers (10 μM) and Phusion polymerase with GC buffer. Both PCR products were inserted into pJV8 using the Golden Gate assembly protocol,

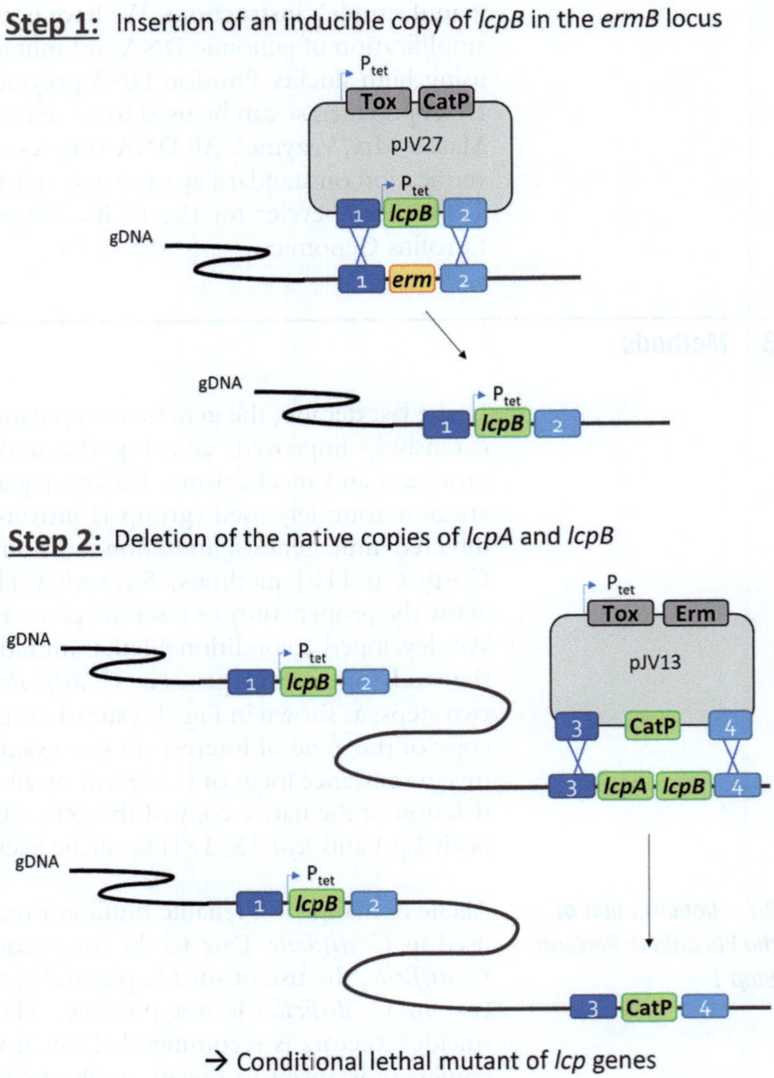

Fig. 1 Global strategy for generating a conditional lethal mutant in *C. difficile*. This strategy is divided into two steps. The first step consists of adding the gene of interest to the *erm* locus. In the second step, the target gene is deleted, and the copy in the *erm* locus is expressed with the help of the P_{tet} promoter. Boxes 1, 2, 3, and 4 represent crossing over parts that are usually 1000 bp in length. Box *erm* represents the erm locus, where two open reading frames encode the genes involved in erythromycin resistance. Tox is the toxin used to help perform the second crossover and CatP is the chloramphenicol resistance cassette

which recommends a similar molarity for each but twofold molarity compared to the plasmid. The reaction mix contained 160 ng of P_{tet}, 178 ng of *lcpB*, and 297 ng of pJV8. The addition of 2.0 μL of T4 DNA ligase buffer 10X, 1 μL of BsaI, 1.5 μL of T4 DNA ligase, and 4 μL of H2O (total volume of reaction mix: 20 μL) allowed the Golden Gate assembly reaction. The enzymatic reaction was

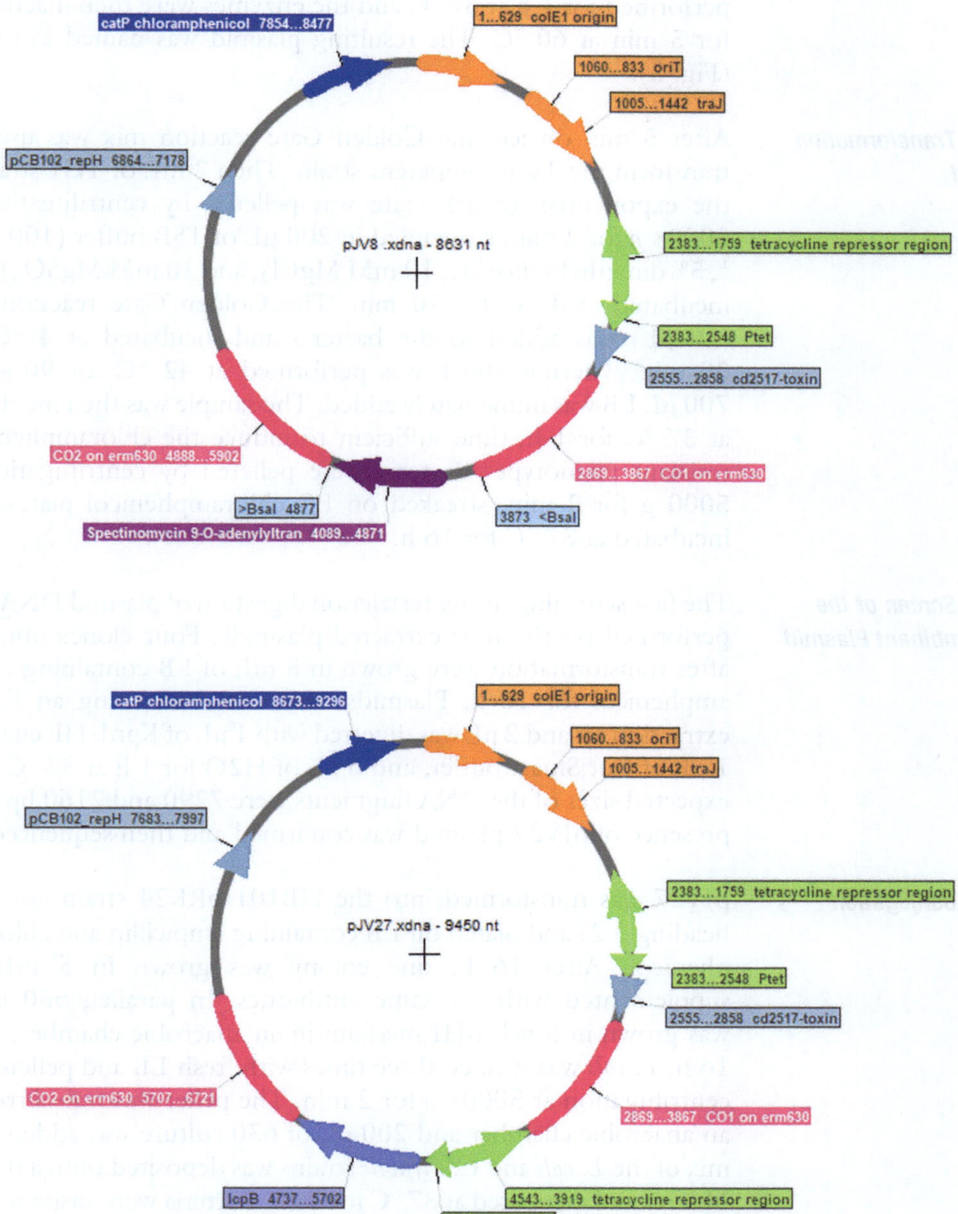

Fig. 2 Graphic maps of the plasmids pJV8 and pJV27 were used in the first step of the protocol. *colE1* is the replication gene for *E. coli*, pCB102_*repH is* the replication gene for *C. difficile*, *oriT* and *traJ* are involved in plasmid conjugation, the tetracycline repressor region and P_{tet} are parts of the P_{tet} promoter, the CD2517 toxin is the toxin that is used to help perform the second crossover, CO1 and CO2 on erm630 are the crossover regions used for the recombination, spectinomycin 9-O-adenyltran is the spectinomycin resistance cassette, and *catP* chloramphenicol is the chloramphenicol resistance cassette. The BsaI restriction sites are indicated

performed for 1 h at 37 °C and the enzymes were then inactivated for 5 min at 60 °C. The resulting plasmid was named as pJV27 (Fig. 2).

3.2 Transformation in TG1

After 5 min on ice, the Golden Gate reaction mix was used to transform the TG1 competent strain. Then 2 mL of TG1 strain in the exponential growth state was pelleted by centrifugation at 5000× *g* for 2 min, suspended in 200 μL of TSB buffer (100 g.L^{-1}, 5% dimethyl sulfoxide, 10 mM $MgCl_2$, and 10 mM $MgSO_4$), and incubated at 4 °C for 30 min. The Golden Gate reaction mix (20 μL) was added to the bacteria and incubated at 4 °C for 30 min. Thermic shock was performed at 42 °C for 90 s and 700 μL LB was immediately added. The sample was then incubated at 37 °C for 1 h, time sufficient to induce the chloramphenicol-resistant phenotype. Bacteria were pelleted by centrifugation at 5000 g for 2 min, streaked on LB–chloramphenicol plates, and incubated at 37 °C for 16 h.

3.3 Screen of the Recombinant Plasmid

The first screening, using restriction digestion of plasmid DNA, was performed on the four extracted plasmids. Four clones obtained after transformation were grown in 5 mL of LB containing chloramphenicol for 16 h. Plasmids were extracted using an EZNA extraction kit and 2 μL was digested with 1 μL of KpnI-HF enzyme, 1 μL of Cut Smart buffer, and 6 μL of H2O for 1 h at 37 °C. The expected sizes of the DNA fragments were 7290 and 2160 bp. The presence of pJV27 plasmid was confirmed and then sequenced.

3.4 Conjugation

pJV27 was transformed into the HB101 pRK24 strain (*see* Subheading 3.2) and plated on LB containing ampicillin and chloramphenicol. After 16 h, one colony was grown in 5 mL LB supplemented with the same antibiotics. In parallel, 630 strain was grown in 5 mL BHI medium in an anaerobic chamber. After 16 h, *E. coli* was washed three times with fresh LB and pelleted by centrifugation at 5000× *g* for 2 min. The pellet was transferred to an anaerobic chamber and 200 μL of 630 culture was added. The mix of the *E. coli* and *C. difficile* strains was deposited onto a 0.22-μM filter and incubated at 37 °C for 16 h. Bacteria were suspended in 1 mL of BHI, and the suspension was spread across 10 plates of BHI containing 16 μg.mL^{-1} of aztreonam and 15 μg.mL^{-1} of thiamphenicol. Colonies were formed after 24–48 h.

3.5 Integration of the lcpB Gene Under the Control of the P_{tet} Promoter

All incubations were performed under anaerobic conditions. Bacterial cells carrying the antibiotic resistance gene *catP* on the pseudo-suicide plasmid (nonintegrated) grow slower in the presence of thiamphenicol because they are limited by the low rate of transmission of this defective plasmid to daughter cells and thus form small colonies. In contrast, the bacteria in which the plasmid was integrated grew faster and produced larger colonies on agar plates supplemented with thiamphenicol, because all daughter cells carried the plasmid. Pure single-crossover integrants can be obtained by selecting and restreaking faster growing colonies and can be validated by PCR screening [14]. One faster growing integrant was streaked on BHI containing 15 μg.mL^{-1} of thiamphenicol. After 24 h, one larger colony was streaked onto a plate of BHI containing 15 μg.mL^{-1} of thiamphenicol. This process was repeated until only large colonies grew on the plate, typically three times. The second crossover event was screened according to the protocol described in ref. [10]. One colony of a pure single crossover integrant was streaked on a plate containing 100 ng.mL^{-1} of ATc and incubated for 16 h at 37 °C. Usually, the presence of a small number of colonies on an ATc plate indicates that the second crossover is successful.

3.6 Screening for Insertion of Inducible lcpB in the ermB Region

Colonies were then screened on BHI containing 5 μg.mL^{-1} of erythromycin, BHI containing 15 μg.mL^{-1} of thiamphenicol, and BHI-only plates until some colonies grew on BHI only. These erythromycin-sensitive colonies were screened by PCR to confirm the insertion of the *lcpB* gene into *ermB* locus (*see* **Note 2**). Two PCRs were performed using the JV99/1998 and 1999/JV100 primers and Phusion polymerase at 55 °C with HF buffer (Fig. 3). The expected amplicon sizes were 2000 bp (*ermB* upstream region of the chromosome and P_{tet}-*lcpB*) and 900 bp (*lcpB-ermB* downstream region of the chromosome). The resulting strain, JMV2, was incubated in BHI for 16 h at 37 °C under anaerobic conditions and frozen at −20 °C.

3.7 Construction of Double lcp (lcpA, lcpB) Deletion Plasmid

To construct the deletion of the targeted gene, in our example, the deletion of the *lcp* locus (containing both *lcpA* and *lcpB*), we used the pJV10 (Fig. 4) plasmid derived from the pJV7 plasmid, which is used to delete genes in chromosome (*see* **Notes 3** and **4**). The pJV10 plasmid is suitable for allelic exchange in an erythromycin-sensitive strain, such as 630Δ*erm* (*see* **Note 4**). The classical 630 Δ*erm* strain still has one *ermB* cassette, which may confer erythromycin resistance. Briefly, three PCR reactions were performed using Phusion polymerase under optimized conditions. The upstream region of *lcpA* and the downstream region of *lcpB* were amplified from the genomic DNA of 630 strains using JV58/JV70 (buffer GC, hybridization temperature 55 °C) and JV73/JV69 primers (buffer GC, hybridization temperature 55 °C). The *catP* cassette was amplified with its promoter from the pMSR plasmid using JV71/JV72 primers (buffer GC, hybridization temperature 55 °C). The sizes of the three PCR products were verified by agarose gel electrophoresis. Each PCR fragment was purified using the EZNA PCR purification kit and quantified using a NanoDrop spectrophotometer.

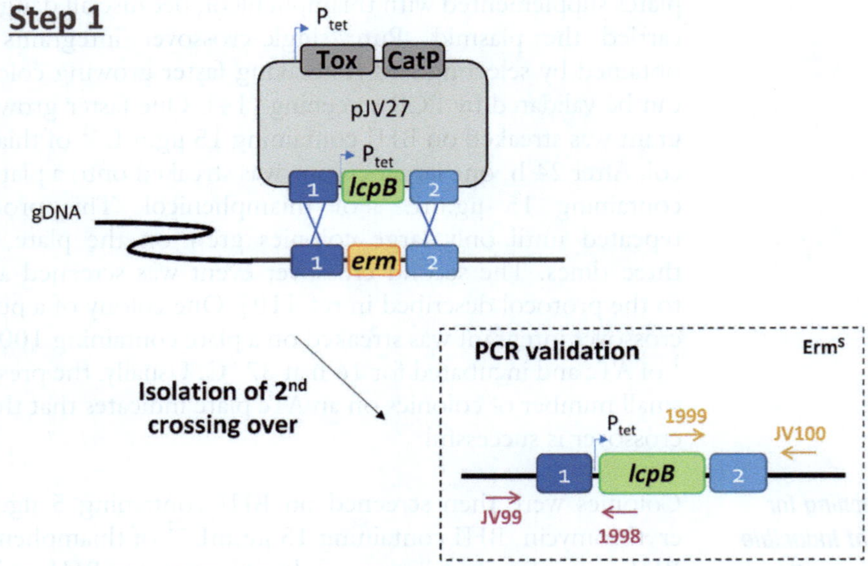

Fig. 3 PCR screening of the insertion of inducible *lcpB* copy in the *ermB* region (**step 1**). Boxes 1 and 2 represent crossing over parts that are usually 1000 bp in length. Box *erm* represents the *erm* locus. Tox is the toxin under the control of the P_{tet} promoter, which is used to help perform the second crossover, and CatP is a chloramphenicol resistance cassette. The P_{tet} promoter allows the expression of *lcpB*. The primer sequences for JV99, 1998, 1999, and JV100 are listed in Table 2

All PCR products were inserted into pJV10 using 243 ng of the upstream region of *lcpA*, 221 ng of the *catP* cassette, 273 ng of the downstream region of *lcpB*, and 373 ng of pJV10 (total insert volume: 15 μL). The addition of 2.5 μL of T4 DNA ligase buffer 10X, 1 μL of BsaI, 1.5 μL of T4 DNA ligase, and 4 μL of H2O (total volume of reaction mix: 25 μL) allowed the Golden Gate assembly reaction. The samples were then incubated for 1 h at 37° C and for 5 min at 60 °C. All Golden Gate samples were transformed into the TG1 strain (*see* Subheading 3.2) and selected on LB containing chloramphenicol and erythromycin. Ten colonies were screened by PCR using JV58/JV69 primers (*see* **Note 5**). Two clones were grown for 16 h at 37 °C in the presence of chloramphenicol and erythromycin. Plasmids were extracted using the EZNA extraction kit, and 2 μL was digested with 1 μL of HindIII enzyme, 1 μL of 2.1 buffer, and 5 μL H2O for 1 h at 37 °C. The expected sizes of the DNA fragments were 5342 bp and 1465 bp. One plasmid with the correct restriction profile was confirmed by sequencing, named pJV13, and transformed into the HB101 pRK24 strain (*see* Subheading 3.2).

3.8 Deletion of Native lcp Locus

Conjugation was performed according to the protocol described in Subheading 3.4. The clones were selected on plates containing aztreonam, thiamphenicol, or erythromycin. The first crossover was determined according to the method described in Subheading 3.6. For the second crossover, as we replaced the ORFs with the

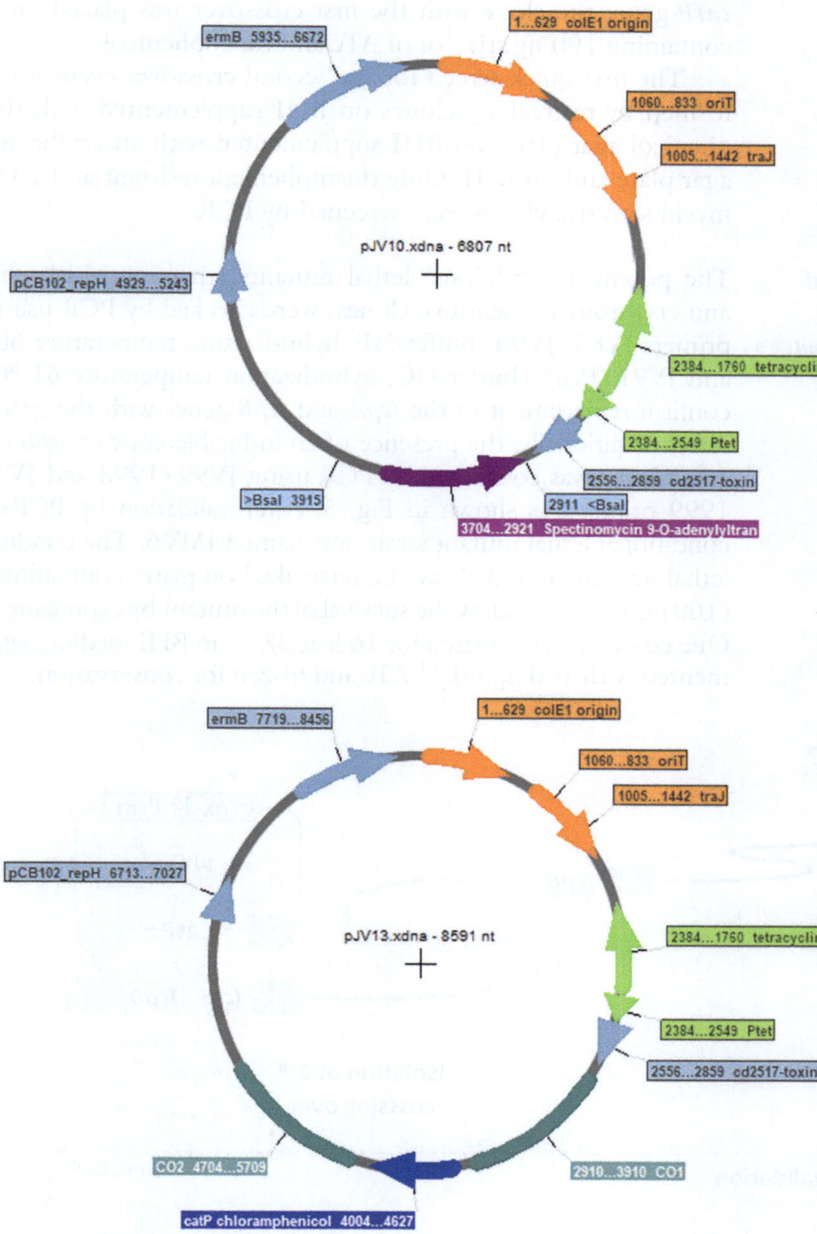

Fig. 4 Graphic maps of plasmids pJV10 and pJV13 used in the second step of the protocol. *colE1* is the replication gene for *E. coli* and pCB102_*repH* the replication gene for *C. difficile*, *oriT* and *traJ* are involved in plasmid conjugation, the tetracycline repressor region and P_{tet} are parts of the P_{tet} promoter, the CD2517 toxin is the toxin that is used to help perform the second crossover, CO1 and CO2 are the crossover regions used for recombination to replace the *lcpA* and *lcpB* genes by the *catP* chloramphenicol resistance cassette, spectinomycin 9-O-adenyltran is the spectinomycin resistance cassette, and *ermB* is the erythromycin resistance cassette. The BsaI restriction sites are indicated

catP gene, the clone with the first crossover was placed on BHI containing 100 ng.uL^{-1}of of ATc and thiamphenicol.

The first quick screen for the second crossover event was performed by restreaking clones on BHI supplemented with thiamphenicol agar plate, on BHI supplemented with an erythromycin agar plate and on BHI. Only thiamphenicol resistant and erythromycin sensitive clones were screened by PCR.

3.9 Screening and Verification of the Obtained Conditional Lethal Mutant

The potential conditional lethal mutants (thiamphenicol-resistant and erythromycin-sensitive clones) were checked by PCR using the primers JV85/JV90 (buffer HF, hybridization temperature 60 °C) and JV91/JV87 (buffer GC, hybridization temperature 61 °C) to confirm replacement of the *lcpA* and *lcpB* genes with the *catP* cassette. Additionally, the presence of an inducible copy of *lcpB* in the *ermB* locus was confirmed by PCR using JV99/1998 and JV100/1999 primers, as shown in Fig. 5. After validation by PCRs, the conditional lethal mutant strain was named JMV6. The conditional lethal mutant should always be restreaked on plates containing ATc (100 ng.mL^{-1}) to allow the survival of the mutant by expressing *lcpB*. One colony was incubated for 16 h at 37 °C in BHI medium supplemented with 100 ng.mL^{-1} ATc and frozen for conservation.

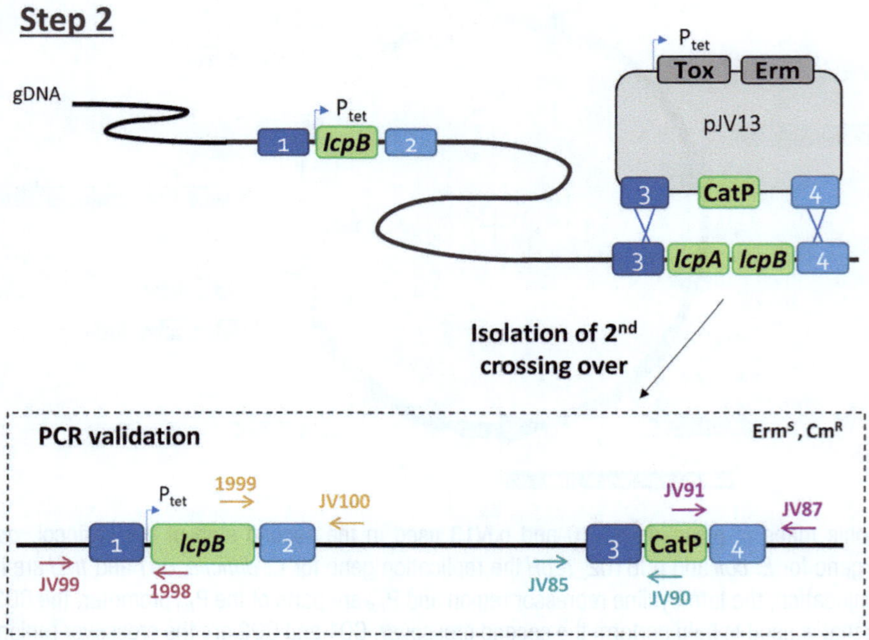

Fig. 5 PCR screening of the deletion of both *lcp* genes (**step 2**). Boxes 3 and 4 represent crossover parts that are usually 1000 bp in length. Box *catP* represents the chloramphenicol resistance cassette with its own promoter. Tox is the toxin under the control of the P_{tet} promoter, which is used to help perform the second crossover, and CatP is a chloramphenicol resistance cassette. The P_{tet} promoter allows the expression of *lcpB*. ErmS and CmR represent clones that were screened by PCR, which are sensitive to erythromycin and resistant to chloramphenicol. The primer sequences for JV99, 1998, 1999, JV100, JV85, JV90, JV91, and JV87 are listed in Table 2

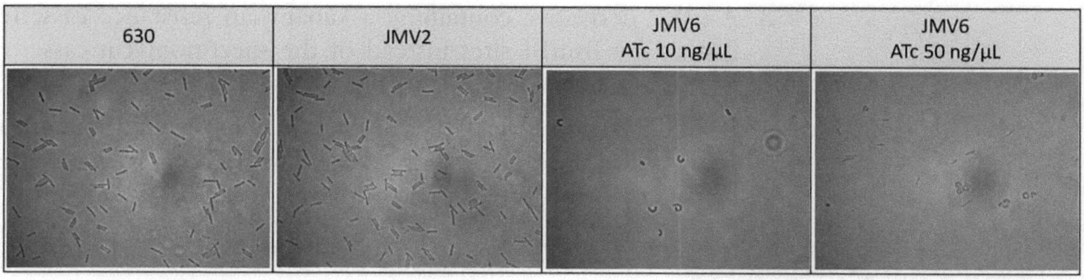

Fig. 6 Modulation of mutant complementation by ATc addition to the medium. The conditional lethal mutant technique allows the modulation of the expression of the gene of interest. This figure shows the complementation of the conditional lethal mutant. 630 and JMV2 strains are control strains, with classical *C. difficile* morphology. In the presence of 10 ng.mL^{-1} of ATc (minimal concentration for mutant survival), the *lcp* conditional lethal mutant (JMV6) loses its rod shape (curved bacilli). When increasing *lcp* expression by increasing ATc concentration to 50 ng.mL^{-1}, JMV6 morphology is restored. Pictures were taken by optical microscopy

3.10 Modulation of the Complementation of the Phenotype by Changing Tet Concentration

In the conditional lethal mutant, JMV6, we controlled the expression of *lcpB* by modifying the concentration of the inducer (ATc) to observe the phenotype of the mutant with low *lcpB* expression. This technique of conditional lethal mutation allows for the observation of phenotypes of strains with low expression of an essential gene of interest and, thus, the understanding of the function of the gene. For gene induction, we tested a wide range of ATc concentrations, ranging from 10 to 250 ng.mL^{-1}[5], two of which are presented in Fig. 6 (the conditional lethal mutant cannot grow without induction of the gene of interest). 630 and JMV2 were used as controls. This technique was the first to study essential genes by modulating their expression in *C. difficile*.

4 Notes

1. It should be noted that if the chosen sequence contained a BsaI site, we also constructed a plasmid with a kanamycin cassette flanked by BsmBI sites instead of BsaI (pMEZ9). Another derivative of pJV8 is the same plasmid but with the chloramphenicol resistance cassette replaced by the erythromycin resistance cassette (pJV17). A pJV17 derivative with BsmBI sites instead of BsaI sites is also available (pMEZ4; *see* Fig. 7).
2. If only erythromycin-resistant colonies appear, restart over to pick two other colonies to perform the first crossover. Indeed, part of the central region between the *erm* genes is homologous to the DNA region upstream of the *erm* locus. If the first crossover occurs in the central region between *erm* genes, after the second crossover, an *erm* gene will remain and colonies will be erythromycin-resistant.

3. A pJV7 derivative containing a kanamycin resistance cassette flanked by BsmBI sites instead of the spectinomycin cassette BsaI is available (pMEZ2; *see* Fig. 7).

4. It should be noted that if the chosen gene to be deleted contains a BsaI site, we also constructed a plasmid with a kanamycin cassette flanked by BsmBI sites instead of BsaI (pMEZ3; *see* Fig. 7).

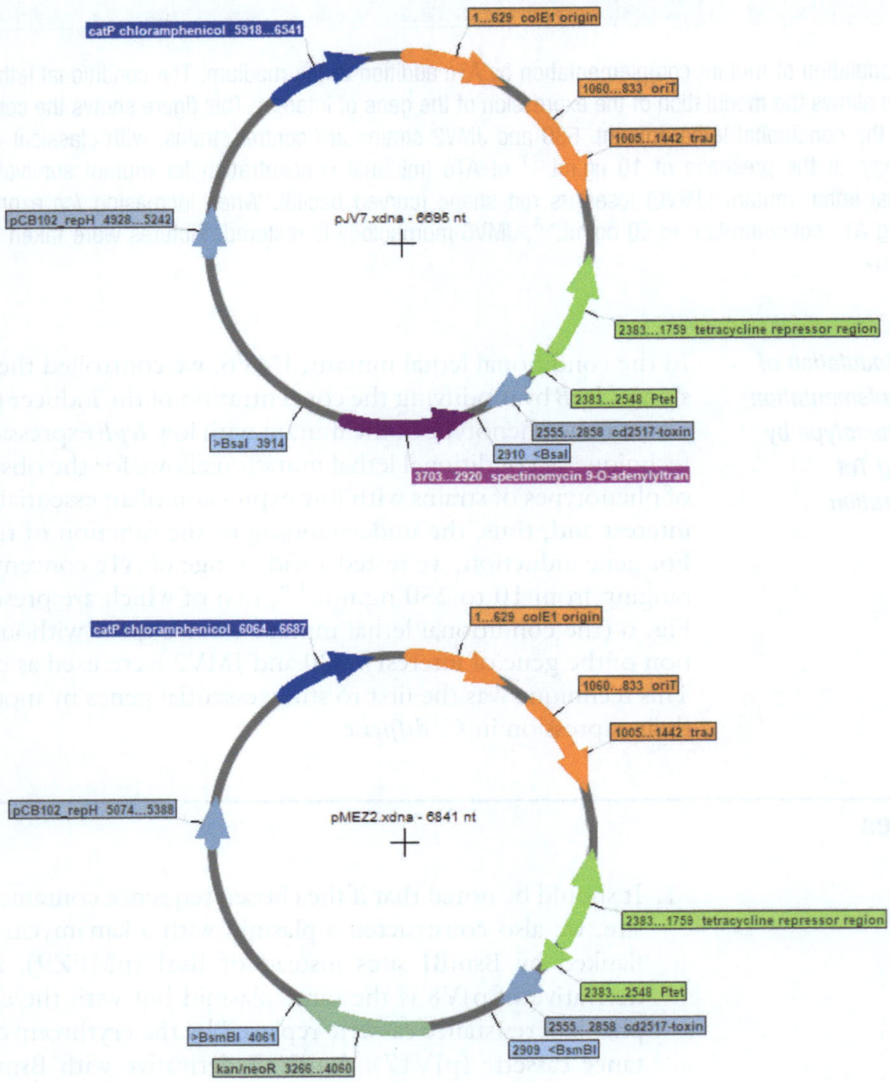

Fig. 7 Graphic maps of other plasmids of the study. pJV7 is a pMSR derivative plasmid used to construct pMEZ2 and pJV10. pMEZ2, pMEZ3, pMEZ4, and pJV17 are used if the chosen sequence contained a BsaI site (plasmids containing BsmBI sites) or if the plasmid must have an *erm* resistance cassette. *colE1* is the replication gene for *E. coli*, pCB102_*repH* the replication gene for *C. difficile*, *oriT* and *traJ* are involved in plasmid conjugation, the tetracycline repressor region and P_{tet} are parts of the P_{tet} promoter, the CD2517 toxin is the toxin that is used to help perform the second crossover, CO1 and CO2 on erm630 are the crossover regions used for the recombination, spectinomycin 9-O-adenyltran is the spectinomycin resistance cassette, *ermB* is the erythromycin resistance cassette, and *catP* chloramphenicol is the chloramphenicol resistance cassette. The BsaI or BsmBI restriction sites are indicated

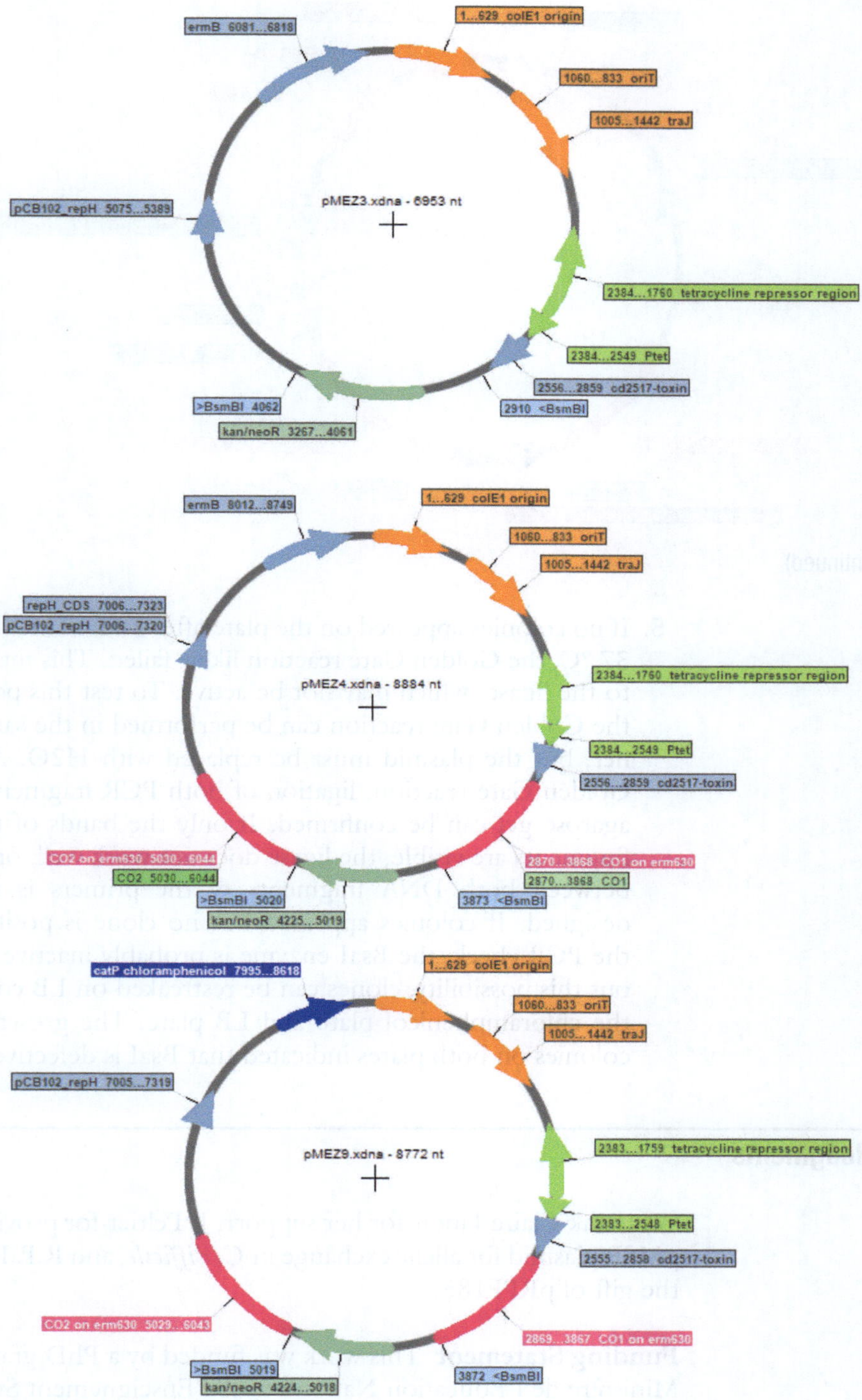

Fig. 7 (continued)

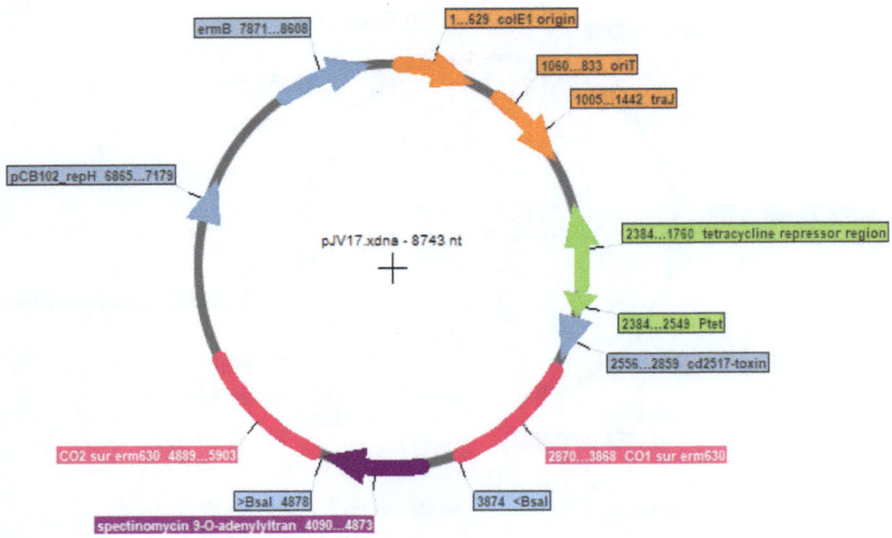

Fig. 7 (continued)

5. If no colonies appeared on the plate after 48 h of incubation at 37 °C, the Golden Gate reaction likely failed. This may be due to the ligase, which may not be active. To test this possibility, the Golden Gate reaction can be performed in the same manner, but the plasmid must be replaced with H2O. After the Golden Gate reaction, ligation of both PCR fragments on an agarose gel can be confirmed. If only the bands of the PCR fragments are visible, the ligase does not work well, or the link between both DNA fragments of the primers is not well designed. If colonies appeared but no clone is positive after the PCR check, the BsaI enzyme is probably inactive. To rule out this possibility, clones can be restreaked on LB containing the chloramphenicol plate and LB plate. The growth of the colonies on both plates indicated that BsaI is defective.

Acknowledgments

We thank Claire Janoir for her support, J. Peltier for providing the pMSR plasmid for allelic exchange in *C. difficile*, and R.P. Fagan for the gift of pRPF185.

Funding Statement This work was funded by a PhD grant of the Ministère de l'Education Nationale, de l'Enseignement Supérieur, de la Recherche et de l'Innovation (MESRI) to J.M.-V. This work was supported by ANR-23-CE18-0037.

References

1. Hegarty JP, Stewart DB (2018) Advances in therapeutic bacterial antisense biotechnology. Appl Microbiol Biotechnol 102:1055–1065
2. Wang B, Kuramitsu HK (2005) Inducible antisense RNA expression in the characterization of gene functions in *Streptococcus mutans*. Infect Immun 73:3568–3576
3. Alberts ME et al. (2025) Analysis of essential genes in *Clostridioides difficile* by CRISPRi and Tn-seq. bioRxiv 2025.06.04.657922. https://doi.org/10.1101/2025.06.04.657922
4. Müh U, Pannullo AG, Weiss DS, Ellermeier CD (2019) A xylose-inducible expression system and a CRISPR interference plasmid for targeted knockdown of gene expression in *Clostridioides difficile*. J Bacteriol 201: e00711–e00718
5. Malet-Villemagne J et al (2023) Polysaccharide II surface anchoring, the Achilles' Heel of *Clostridioides difficile*. Microbiol Spectrum 11: e0422722
6. Alexiou S, Diakou A, Kachrimanidou M (2025) The role of *Clostridioides difficile* within the one health framework: a review. Microorganisms 13:429
7. Morado F, Nanda N (2024) A review of therapies for *Clostridioides difficile* infection. Antibiotics (Basel) 14:17
8. Zvonareva T, Courson DS, Purcell EB (2025) *Clostridioides difficile* infection study models and prospectives for probing the microbe-host interface. J Bacteriol 207:e0040724
9. Fagan RP, Fairweather NF (2011) *Clostridium difficile* has two parallel and essential Sec secretion systems. J Biol Chem 286:27483–27493
10. Peltier J et al (2020) Type I toxin-antitoxin systems contribute to the maintenance of mobile genetic elements in *Clostridioides difficile*. Commun Biol 3:718
11. Sebaihia M et al (2006) The multidrug-resistant human pathogen *Clostridium difficile* has a highly mobile, mosaic genome. Nat Genet 38:779–786
12. Kuehne SA, Rood JI, Lyras D (2019) Clostridial genetics: genetic manipulation of the pathogenic clostridia. Microbiol Spectrum 7
13. Chu M et al (2016) A *Clostridium difficile* cell wall glycopolymer locus influences bacterial shape, Polysaccharide production and virulence. PLoS Pathog 12:e1005946
14. Cartman ST, Kelly ML, Heeg D, Heap JT, Minton NP (2012) Precise manipulation of the *Clostridium difficile* chromosome reveals a lack of association between the *tcdC* genotype and toxin production. Appl Environ Microbiol 78:4683–4690

Chapter 4

CRISPR-AsCas12a and dAsCas12a-Mediated Gene Knockout and Knockdown in *Clostridioides difficile*

Duolong Zhu, Shaohui Wang, Xingmin Sun, and Robert A. Britton

Abstract

Clostridioides difficile (*C. difficile*) is a leading cause of antibiotic-associated diarrhea and severe colitis, yet its genetic manipulation has long been constrained by low DNA transfer efficiency and limited recombination systems. Recent advances in CRISPR-based technologies have revolutionized the genetic toolkit for this pathogen, enabling precise genome editing and transcriptional regulation. Among CRISPR nucleases, Cas12a offers distinct advantages over Cas9 for bacterial applications, including a smaller size, T-rich PAM recognition, single-crRNA requirement, and reduced toxicity, which enhances conjugation efficiency in genetically recalcitrant organisms. AsCas12a-based platforms have enabled large fragment deletions, multiplex editing, and rapid generation of marker-free mutants in *C. difficile*. Complementing these nuclease-active systems, nuclease-deactivated variants (dCas9 or dAsCas12a) support CRISPR interference (CRISPRi)—a reversible, tunable approach for transcriptional repression without altering genomic sequences. Compared to traditional mutagenesis, CRISPRi greatly accelerates functional genomics by enabling high-throughput screening and drug target discovery. Together, our lab has independently developed CRISPR-AsCas12a-mediated genome editing and dAsCas12a-based CRISPRi tools, providing complementary strategies to overcome longstanding genetic barriers in *C. difficile*. These tools open new avenues for system-level interrogation of virulence, antibiotic resistance, and host-pathogen interactions.

Key words CRISPR, CRISPRi, AsCas12a, dAsCas12a

1 Introduction

Clostridioides difficile (*C. difficile*) is a leading cause of antibiotic-associated diarrhea and life-threatening colitis, yet its genetic manipulation has long been challenging due to low DNA transfer efficiency and limited recombination systems [1, 2]. Recent advances in CRISPR-based technologies have revolutionized the genetic toolbox for *C. difficile*, enabling precise genome editing and transcriptional regulation [3–5].

CRISPR-Cas systems, originally discovered as adaptive immune mechanisms in prokaryotes, have been repurposed for targeted gene editing and gene expression control [6]. While

Kurt H. Piepenbrink and Jennifer M. Auchtung (eds.), *Clostridioides difficile: Methods and Protocols*, Methods in Molecular Biology, vol. 3046, https://doi.org/10.1007/978-1-0716-5328-9_4,

Cas9 (type II system) was the first widely adopted nuclease, Cas12a (Cpf1) (type V system) offers distinct advantages for bacterial applications [7]. Unlike Cas9, Cas12a requires only a single crRNA (no tracrRNA), recognizes a T-rich PAM (5′-TTTV-3′), and introduces staggered double-strand breaks distal to the PAM site. Cas12a is smaller (~151 kDa vs. ~163 kDa for SpCas9), facilitates streamlined plasmid design, and reduces off-target effects. Importantly, Cas12a exhibits lower toxicity than Cas9, improving conjugation efficiency in genetically recalcitrant organisms such as *C. difficile*. Its architecture also simplifies multiplex genome editing using crRNA arrays. Using inducible systems, AsCas12a has been successfully applied for large-fragment deletions (up to 49 kb), multiplex editing, and rapid construction of clean mutants without antibiotic markers [4].

In addition to nuclease-active CRISPR systems for gene knockout, nuclease-deactivated variants (dCas9 or dCas12a) enable programmable transcriptional repression without altering genomic sequences. This approach, known as CRISPR interference (CRISPRi), blocks RNA polymerase progression at targeted loci, offering a reversible and tunable strategy to study essential genes and operons [8]. In *C. difficile*, xylose-inducible CRISPRi platforms have achieved up to 100-fold knockdown of essential genes such as *ftsZ*, *slpA*, and *pbp-0712*, confirming their roles in cell division and envelope integrity [9]. Recently, CRISPRi was also employed to identify essential genes in *C. difficile* strain R20291 and compared with previous transposon insertion site sequencing (Tn-seq) [10, 11]. Of the 404 putatively essential genes previously identified, 181 were targeted using CRISPRi, confirming essentiality for over 90% of them.

We have developed a CRISPR-AsCas12a-mediated genome editing tool for *C. difficile* [12] and adapted this system into a CRISPR-dAsCas12a platform by introducing a single amino acid substitution (E993A) to abolish DNA cleavage activity [13]. This modified system was successfully applied to knock down the toxin B gene (*tcdB*) in R20291 (Fig. 1). Together, CRISPR-AsCas12a-based genome editing and dAsCas12a-mediated gene knockdown offer complementary approaches for functional genetic analysis in *C. difficile*. These tools overcome longstanding genetic barriers and enable system-level interrogation of virulence, antibiotic resistance, and host-pathogen interactions.

2 Materials

1. Bacteria: *Escherichia coli* DH5α (hereafter DH5α) was used as the cloning host. *E. coli* HB101 carrying pRK24 (hereafter HB101) served as the conjugation donor. *Clostridioides difficile* was used as the target strain for gene editing.

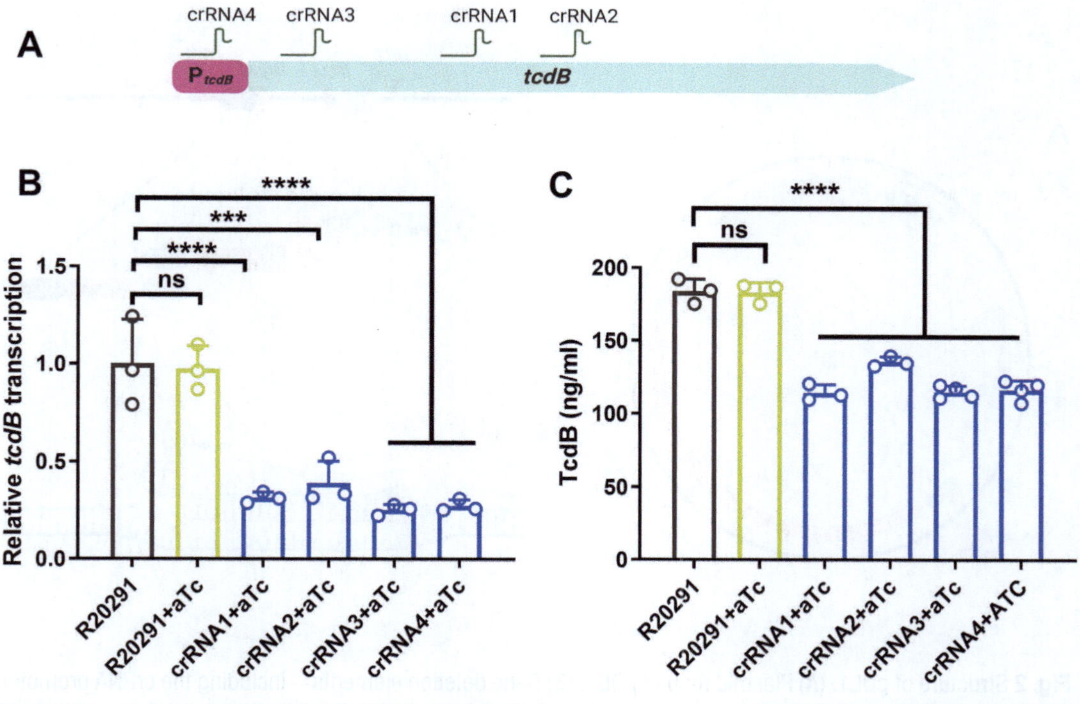

Fig. 1 Example of *tcdB* gene knockdown in *C. difficile* R20291. (**A**) Four distinct crRNAs targeting different regions of the *tcdB* gene. (**B**) Relative *tcdB* mRNA levels measured by RT-qPCR. (**C**) TcdB concentration in the culture supernatants quantified by ELISA. The inducer anhydrotetracycline (500 ng/mL) was added to late log-phase R20291 cultures and incubated for 4 h before harvesting cells for RNA extraction and collecting culture supernatants for ELISA analysis

2. Plasmids: pDL1 (pMTL82151/P_{tet}-AsCas12a) for gene deletion (Fig. 2); pDL2 for gene knockdown in *C. difficile* (Fig. 3) (*see* **Note 1**).
3. Media: LB medium (1% tryptone, 0.5% yeast extract, 1% NaCl) was used for *E. coli* cultures. BHIS medium (Difco™ brain–heart infusion broth supplemented with 0.5% yeast extract and 0.1% L-cysteine) was used for *C. difficile* cultures. Agar plates were prepared by adding 1.5% agar to the respective media.
4. Antibiotics and selection media: For DH5α carrying pDL plasmids (gene deletion plasmid pDL1 or gene knockdown plasmid pDL2), LB media supplemented with 15 μg/mL chloramphenicol (LB-C_{15}) was used. For HB101 carrying pDL plasmids, LB media containing 15 μg/mL chloramphenicol and 100 μg/mL ampicillin (LB-$C_{15}A_{100}$) was used. For *C. difficile* conjugation selection, BHIS agar plates were supplemented with 15 μg/mL thiamphenicol, 50 μg/mL kanamycin, and 8 μg/mL cefoxitin (BHIS-TKC). Maintenance of pDL plasmids in *C. difficile* was performed in BHIS media containing 15 μg/mL thiamphenicol (BHIS-T_{15}). Gene deletion

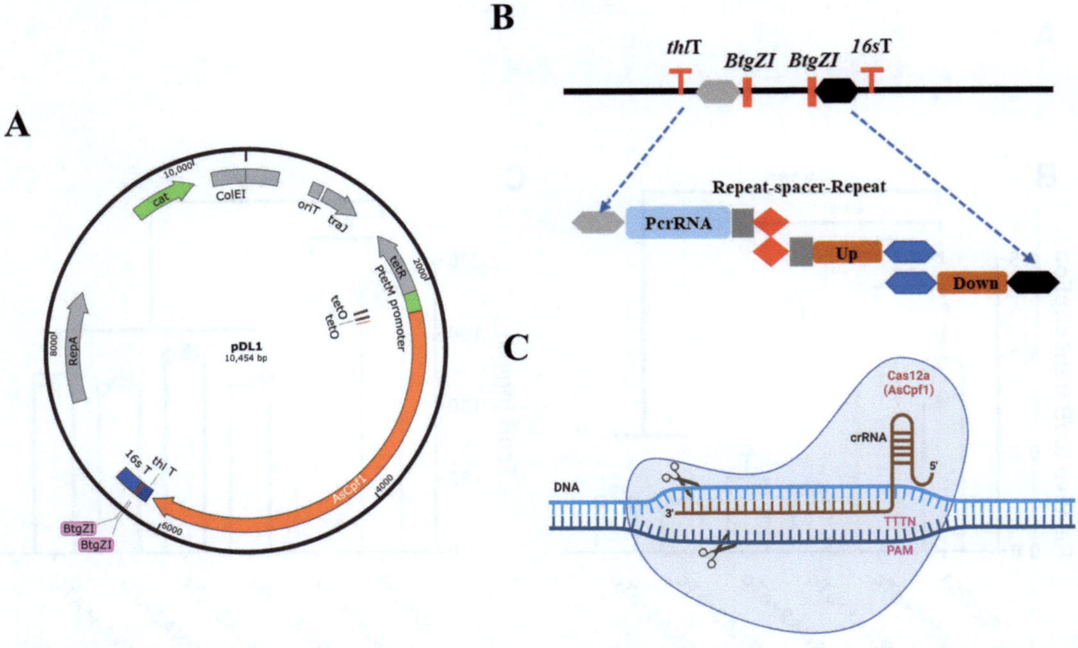

Fig. 2 Structure of pDL1. (**A**) Plasmid map of pDL1. (**B**) Gene deletion elements—including the crRNA promoter (PcrRNA), repeat–spacer–repeat sequence, and upstream/downstream homology arms—were assembled into *Btg*ZI-digested pDL1. The PcrRNA is derived from the promoter of the small RNA gene *sCbei_5830* ([3]). (**C**) Sketch of AsCas12a-mediated genomic DNA cleavage

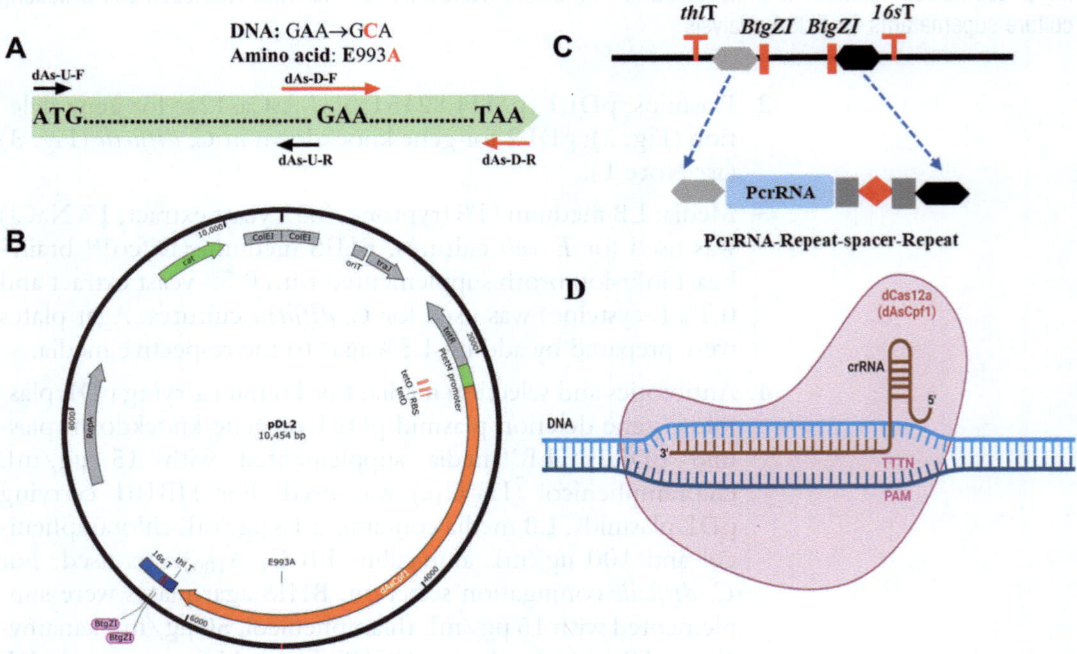

Fig. 3 Structure of pDL2. (**A**) The key amino acid residue of AsCas12a in pDL1 was mutated (E993A) by overlap PCR to generate the catalytically inactive variant, dAsCas12a; (**B**) plasmid map of pDL2; (**C**) gene knockdown cassette (PcrRNA-repeat–spacer–repeat) was assembled into *Btg*ZI-digested pDL2; (**D**) Sketch of dAsCas12a-mediated gene knockdown

induction in *C. difficile* was achieved using BHIS media or agar plates supplemented with 500 ng/mL anhydrotetracycline (BHIS-aTc).

5. DNA primers and biology reagents: All DNA primers were synthesized by Integrated DNA Technologies (IDT). Homology arms for target gene editing were designed to be 800–1000 bp in length. Restriction enzyme *Btg*ZI, NEB Q5 High-Fidelity DNA Polymerase, NEBuilder HiFi DNA Assembly Master Mix, and DNA gel extraction kits were purchased from New England Biolabs (NEB).

3 Methods

3.1 Design Gene-Editing Plasmid

1. Guide RNA (gRNA) design: Multiple gRNAs targeting distinct regions of the interested gene were designed using CRISPR RGEN Tools (http://www.rgenome.net/cas-designer/) and evaluated for potential off-target effects using Cas-OFFinder (http://www.rgenome.net/cas-offinder/).
2. Amplify crRNA promoter: For crRNA expression, the promoter (PcrRNA) used in plasmid pDL1 was the same as that in plasmid pWH34 [4]. A synthetic PcrRNA (*see* **Note 2**) containing one crRNA repeat sequence (TAATTTC TACTCTTGTAGAT) was ordered from IDT. This PcrRNA was PCR-amplified for each assembly using a universal forward primer (Uni-F: AAAGTTAAAAGAAGAAAATAGAAATA TAATCTTTAATTTGAAAAGATTTA) and a reverse primer (R: rN_{23}ATCTACAAGAGTAGAAATTAATGGT). For gene knockdown constructs, a modified reverse primer (R: CATG CTGATCTAGATTTCTCCATAGATCTACAAGAGTA GAAATTArN_{23}ATCTACAAGAGTAGAAATTAATGGT) was used to generate a promoter–repeat–crRNA–repeat cassette, which was assembled into plasmid pDL2 (*see* **Note 3**).
3. Homology arm selection: 800–1000 bp upstream and downstream of the target gene was PCR amplified for recombination. The forward primer for the upstream arm included a 25-nucleotide overhang (ATN_{23}) at the 5′ end to facilitate assembly with the PcrRNA region. The forward primer for the downstream arm included a 25-nucleotide overlapping with the upstream arm, while the reverse primer carried a 25-nucleotide overhang (CATGCTGATCTAGATTTCTCCA TAG) to aid in plasmid assembly (*see* **Note 4**).
4. Assembly: The crRNA promoter (PcrRNA) containing a repeat and spacer sequence (PcrRNA–repeat–spacer), along with the homology arms, was assembled into plasmid pDL1 for gene deletion. For gene knockdown, a knockdown crRNA element

(PcrRNA–repeat–spacer–repeat) was assembled into plasmid pDL2. All assemblies were performed using the NEBuilder HiFi DNA Assembly Master Mix in a total reaction volume of 10 μL, incubated at 50 °C for 1 h in a thermal cycler, following the manufacturer's instructions.

3.2 Transform Gene-Editing Plasmid to E. coli

The assembly mix from Subheading 3.1, **Step 4** was transformed into DH5α and plated on LB-C_{15} agar plates. Colonies were screened by colony PCR using the pDL1 check primer pairs (pDL1-C-F: TACGAATTCGAGCTCGGTACCCGGG; pDL1-C-R: CGCGTGACGTCGACTCTAGAGGATC). Expected PCR product sizes for the correct plasmid: ~2.7 kb (includes ~2 kb homologous arms) and the empty plasmid is about 410 bp. The verified plasmid was amplified and purified from DH5α and then fully sequenced. The confirmed plasmid was subsequently transformed into HB101.

3.3 Conjugate Gene-Editing Plasmid into C. difficile

1. Inoculate a single colony of HB101 carrying the target plasmid into 5 mL LB-C_{15} media and culture aerobically at 37 °C overnight. Simultaneously, inoculate a fresh colony of the target *C. difficile* strain into 5 mL BHIS media (pre-reduced) and incubate anaerobically at 37 °C overnight.
2. Next morning, centrifuge 2–3 mL of HB101 cultures at 4000 × *g* for 2 min, wash once with BHIS media, and centrifuge again. Transfer the bacterial pellets into the anaerobic chamber. Add 200–300 μL of overnight *C. difficile* cultures to the HB101 pellets and gently resuspend.
3. Spot 40–50 μL of the resuspended mixtures onto BHIS plates and incubate anaerobically at 37 °C for 12–18 h (not exceeding 24 h).
4. Scrape the bacterial spots from the mating plates into 1-mL sterile PBS using a 10-μL sterile inoculation loop, resuspend thoroughly, and then spread 100 μL onto each of ten BHIS-TKC plates. Incubate anaerobically at 37 °C until distinct single transconjugant colonies are visible.
5. Inoculate a single transconjugant into BHIS-T_{15} media and prepare a glycerol stock of *C. difficile* transconjugant.

3.4 Induction for Gene Deletion in C. difficile

1. Grow the single transconjugant in BHIS-T_{15} media to $OD_{600nm} \approx 0.8$. Following, induce AsCas12a expression by either streaking on BHIS-T_{15}-aTc_{500} plates, or plating 100 μL of ten-fold dilutions (10^4 ~ 10^7) on BHIS-T_{15}-aTc_{500} plates.
2. Incubate plates anaerobically at 37 °C until colonies are large enough for PCR screening using the target gene specific check primer pairs (*see* **Note 5**). Sequencing PCR products to confirm the correct deletion.

3.5 Plasmid Curing

1. Subculture confirmed mutants in antibiotic-free BHIS for 10–15 passages to cure the plasmid.
2. Serially dilute cultures in PBS, plate 100 μL of diluted cultures (10^4–10^7) on BHIS plates, and incubate anaerobically at 37 °C.
3. Screen colonies by streaking on BHIS and BHIS-T_{15} plates. Colonies growing only on BHIS plates are plasmid cured mutants (*see* **Note 6**).
4. *Optional:* Extract genomic DNA and confirm by whole-genome sequencing.

3.6 Induction for Gene Knockdown in C. difficile

1. Follow the same conjugation steps described in Subheading 3.3 to obtain *C. difficile* transconjugants harboring the gene knockdown plasmid.
2. Add anhydrotetracycline (100–500 ng/mL) to cultures at early or mid-log phase to induce expression of the knockdown construct. (*see* **Note** 7).
3. After 2–4 h of induction, isolate bacterial total RNA and assess target gene expression using RT-qPCR. If phenotypic changes are anticipated (e.g., toxin production, biofilm formation, or fluorescence), knockdown efficiency can be further validated by ELISA for toxin quantification, biofilm assays for biomass measurement, and fluorescence intensity analysis with a plate reader.

4 Notes

1. We developed our own CRISPR-AsCas12a-mediated genome editing plasmid (pDL1) and dAsCas12a-based gene knockdown plasmid (pDL2). AsCas12a (AsCpf1), synthesized by IDT, was placed under the control of the P_{tet} promoter, incorporating tetracycline-inducible elements from pRPF185. To enable modular crRNA cloning and expression, two *Btg*ZI restriction sites flanked by the thlT and 16sT terminators (thlT–*Btg*ZI–*Btg*ZI–16sT) were integrated into the pMTL82151 backbone. The key amino acid residue of AsCas12a in pDL1 was mutated (E993A) to generate the catalytically inactive variant, dAsCas12a, used in plasmid pDL2. The plasmids can be requested from either Dr. Duolong Zhu (duolong.zhu@bcm.edu) or Dr. Xingmin Sun (sun5.usf.edu).
2. crRNA promoter (PcrRNA) with one *crRNA repeat*
   ATAATCTTTAATTTGAAAAGATTTAAGGCTTATT-TAAATAAAAAATATGAGGGAAGAATTGATATAAATT-TAATTTTGTTATTGTATTATGGTATGTATGGAA-

TAAATTTAACATAAAGACAGTAATAATGTTCTTGAATT-TAGACTTTTTATGTGTTATCATTAACAAGTAT-CAAAAATGACATTTAATAAATTAATAATAATTTTAAAAA-TATATTTTTGATAAAAGCAATGATTAACATGGTTT-GACGTCTGAGAAGAGACGATTTTCTCAATAGGA-GAAATTAAGGTGCAAACCCTTATCATTCCACCAT *TAATTTCTACTCTTGTAGAT*

3. *r*N_{23}: reverse complementation of designed 23-nucleotide gRNA sequence (spacer).
4. N_{23}: designed 23-nucleotide gRNA sequence (spacer).
5. Check primers were typically designed to anneal 50–100 bp away from the ends of the upstream and downstream homology arms. In certain *C. difficile* ribotypes (e.g., RT078 and RT012), colony PCR may produce both wild-type and correct gene deletion bands, indicating that a single colony contains a mixture of wild-type and mutant cells. Restreaking the selected colony onto BHIS-T_{15}-aTc_{500} plates can improve purity and yield a fully deleted mutant. For some strains where restreaking does not resolve mixed populations, replacing the crRNA expression promoter with the *C. difficile tuf* promoter (*C. difficile* M120 *tuf* promoter: TTTAAAAATAATTAATAA TATAAATATAAAAAATAATCAGCTAGATGAGATATATCT CATCTAGCACAAGATAAAGGGAGGAAATTCAAA) can enhance gene deletion efficiency and facilitate isolation of pure mutants. Both pDL1 and pDL2 perform reliably in strain R20291, typically producing a single PCR band—either wild type or mutant.
6. For unknown reasons, some target gene deletion plasmids are difficult to cure, even after 30 to 50 passages. Culturing the deletion mutant in BHIS medium supplemented with 500 ng/mL anhydrotetracycline—without the plasmid selection antibiotic thiamphenicol—can significantly improve plasmid curing efficiency.
7. The concentration of anhydrotetracycline and the bacterial growth phase at the time of induction should be optimized based on the specific experimental design and target gene expression dynamics.

References

1. Zhu D, Sorg JA, Sun X (2018) Clostridioides difficile biology: sporulation, germination, and corresponding therapies for C. difficile infection. Front Cell Infect Microbiol 8:29
2. Minton N, Carter G, Herbert M, O'Keeffe T, Purdy D, Elmore M, Ostrowski A, Pennington O, Davis I (2004) The development of Clostridium difficile genetic systems. Anaerobe 10(2):75–84
3. McAllister KN, Bouillaut L, Kahn JN, Self WT, Sorg JA (2017) Using CRISPR-Cas9-mediated genome editing to generate C. difficile mutants defective in selenoproteins synthesis. Sci Rep 7(1):14672

4. Hong W, Zhang J, Cui G, Wang L, Wang Y (2018) Multiplexed CRISPR-Cpf1-mediated genome editing in Clostridium difficile toward the understanding of pathogenesis of C. difficile infection. ACS Synth Biol 7(6): 1588–1600
5. Shen A (2019) Expanding the Clostridioides difficile genetics toolbox. J Bacteriol 201(14)
6. Marraffini LA (2015) CRISPR-Cas immunity in prokaryotes. Nature 526(7571):55–61
7. Swarts DC, Jinek M (2018) Cas9 versus Cas12a/Cpf1: structure-function comparisons and implications for genome editing. Wiley Interdiscip Rev: RNA 9(5):e1481
8. Larson MH, Gilbert LA, Wang X, Lim WA, Weissman JS, Qi LS (2013) CRISPR interference (CRISPRi) for sequence-specific control of gene expression. Nat Protoc 8(11): 2180–2196
9. Muh U, Pannullo AG, Weiss DS, Ellermeier CD (2019) A xylose-inducible expression system and a CRISPR interference plasmid for targeted knockdown of gene expression in Clostridioides difficile. J Bacteriol 201(14)
10. Alberts ME, Kurtz MP, Muh U, Bernardi JP, Bollinger KW, Dobrila HA, Duncan L, Laster HM, Orea AJ, Pannullo AG, Rivera-Rosado JG, Torres FV, Ellermeier CD, Weiss DS (2025) Analysis of essential genes in Clostridioides difficile by CRISPRi and Tn-seq. J Bacteriol 207(10):e0022025
11. Dembek M, Barquist L, Boinett CJ, Cain AK, Mayho M, Lawley TD, Fairweather NF, Fagan RP (2015) High-throughput analysis of gene essentiality and sporulation in Clostridium difficile. mBio 6(2):e02383
12. Zhu D, Patabendige H, Tomlinson BR, Wang S, Hussain S, Flores D, He Y, Shaw LN, Sun X (2021) Cwl0971, a novel peptidoglycan hydrolase, plays pleiotropic roles in Clostridioides difficile R20291. Environ Microbiol 23(9):5222–5238
13. Yamano T, Nishimasu H, Zetsche B, Hirano H, Slaymaker IM, Li Y, Fedorova I, Nakane T, Makarova KS, Koonin EV, Ishitani R, Zhang F, Nureki O (2016) Crystal structure of Cpf1 in complex with guide RNA and target DNA. Cell 165(4):949–962

Chapter 5

Isolation and Characterization of *Clostridioides difficile* Bacteriophages

Alexia L. M. Royer, Andrew A. Umansky, and Louis-Charles Fortier

Abstract

Bacteriophages, or simply phages, are viruses that infect bacteria. They are the most abundant biological entity on Earth and their impact on the physiology and evolution of bacteria is tremendous. Phages have been at the forefront of developments in molecular biology and genetics. With the rise in antibiotic resistance worldwide, they are now also viewed as powerful antimicrobials to fight multidrug-resistant pathogens. Several methods have been developed over decades of fundamental phage research to work with phages. Although some of these methods are readily applicable to most phages, others need to be specifically adapted to a given phage–host ecosystem. For example, *Clostridioides difficile* is a strict anaerobe and methods that are normally performed on the open bench need to be done inside an anaerobic chamber, which complicates working with phages. Phage propagation in liquid and on solid agar can also be influenced by different parameters such as the concentration of agar or agarose and the presence of divalent cations. Our lab has 20 years of experience working with *C. difficile* phages and this chapter describes basic methods that we use to isolate, purify, propagate, and characterize phages infecting *C. difficile*.

Key words Bacteriophage isolation and propagation, Prophage induction, Prophage prediction tool

1 Introduction

Bacteriophages targeting *Clostridioides difficile* have been the subject of growing scientific interest for over two decades. Numerous phages infecting *C. difficile* have been isolated, characterized, and sequenced over the years, and despite significant advances, their functional roles in bacterial physiology, virulence, and ecology remain poorly understood [1]. Initially employed as a typing tool, their relevance expanded significantly following the publication of the first complete phage genome (phiCD119) in 2006 [2]. This milestone coincided with major *C. difficile* outbreaks in North America and Europe, prompting exploration of phages as potential alternatives to antibiotics. Indeed, their high specificity would offer a distinct advantage: the ability to selectively eliminate *C. difficile*

Kurt H. Piepenbrink and Jennifer M. Auchtung (eds.), *Clostridioides difficile: Methods and Protocols*, Methods in Molecular Biology, vol. 3046, https://doi.org/10.1007/978-1-0716-5328-9_5,

from the gut microbiota without disrupting commensal bacterial populations.

Notably, all characterized *C. difficile* phages to date are temperate, which is a significant limitation for therapeutic applications. Given the risks associated with lysogeny—such as horizontal gene transfer and lysogenic conversion—strictly lytic phages are generally preferred for clinical use. Recent developments in genetic engineering have opened new avenues to overcome this limitation. With the increasing accessibility of molecular tools for manipulating *C. difficile*, it is now feasible to genetically modify temperate phages in their prophage state. This includes converting them into obligately lytic forms and engineering tail fiber proteins to expand or alter host range, thereby enhancing their therapeutic potential [3].

Working with *C. difficile* phages presents unique technical challenges, primarily due to the obligate anaerobic nature of the host bacterium. Although phages themselves are not oxygen-sensitive, prolonged exposure of *C. difficile* to oxygen can compromise experimental outcomes. Consequently, many procedures, including soft agar overlay preparation and certain centrifugation steps, must be conducted within an anaerobic chamber.

While the core methodologies for phage isolation and propagation have remained relatively unchanged, significant progress has been made in the development of molecular and bioinformatic tools to manipulate and analyze *C. difficile* phages. These include advanced pipelines for prophage identification, genome annotation, and comparative genomics, which have greatly enhanced our ability to study phage diversity and function.

This chapter outlines current strategies for identifying prophages within *C. difficile* genomes, inducing them from lysogenic strains and propagating them on susceptible hosts. Many of these methodological steps can be used to isolate free phages from environmental sources (e.g., feces, soil, wastewater, etc.). We also briefly describe protocols for phage characterization, including adsorption and infection assays, whole genome sequencing, and transmission electron microscopy. Together, these approaches provide a comprehensive framework for advancing our understanding of *C. difficile* phages and harnessing their potential for therapeutic applications.

2 Materials

2.1 Prophage Induction with Mitomycin C

1. Pre-reduced TY (3% tryptose, 2% yeast extract, pH 7.4) liquid broth.
2. Sterile borosilicate glass tubes with caps suitable for optical density measurements (16 × 150 mm, Fisher Scientific, Cat. No.14-961-31).

3. Spectrophotometer set to a 600-nm wavelength (OD_{600}) with an adapter for glass tubes or a portable tube densitometer.
4. Mitomycin C (Fisher Scientific, Cat. No. NC1081159). Prepare a 10 mg/mL stock solution in DMSO and dilute in sterile distilled water to a 1 mg/mL working solution. Store at −20° C. Avoid repeated freeze-thaw cycles and use of old solutions as they can lose their activity and lead to unreliable induction.
5. Conical 15-mL polypropylene centrifuge tubes (Sarstedt, Cat. No. 62.554.205, or equivalent).
6. Refrigerated centrifuge equipped with appropriate rotor for the spinning of 15-mL tube at ≥3000 × *g*.
7. Disposable 10-mL syringes (Fisher Scientific, Cat. No. 14-823-2A, or equivalent).
8. Disposable 0.22-μm or 0.45-μm syringe filters (Sarstedt, Cat. No. 83.1826.001, 83.1826, or equivalent).

2.2 Prophage Induction in Broth and on Agar Plates Using Ultraviolet Light

1. Pre-reduced TY liquid broth (3% tryptose, 2% yeast extract, pH 7.4).
2. Pre-reduced TY + 1% agar plates (10 g/L agar in TY).
3. Sterile borosilicate glass tubes with caps suitable for optical density measurements (16 × 150 mm, Fisher Scientific, Cat. No.14-961-31).
4. Spectrophotometer set to a 600-nm wavelength (OD_{600}) with an adapter for glass tube or a portable tube densitometer.
5. Conical 15-mL polypropylene centrifuge tubes (Sarstedt, Cat. No. 62.554.205, or equivalent).
6. Refrigerated centrifuge equipped with appropriate rotor for the spinning of 15 mL tube at ≥3000 × *g*.
7. Sterile 3-mm or 6-mm glass beads (Fisher Scientific, Cat. No. CG110102, 11-312D, or equivalent) or inoculation spreader (Sarstedt, Cat. No. 86.1569.005 or equivalent).
8. UV transilluminator (wavelength: 302–365 nm). Conventional gel documentation transilluminators are suitable for this application.
9. Disposable 10-mL serological pipettes (Sarstedt, Cat. No. 86.1254.001).
10. Disposable 10-mL syringes (Fisher Scientific, Cat. No.14-823-2A, or equivalent).
11. Disposable 0.22-μm or 0.45-μm syringe filters (Sarstedt, Cat. No. 83.1826.001, 83.1826, or equivalent).
12. TY soft agarose (0.3 g/L agarose in TY) melted and kept at 55 °C in a block heater.
13. Dry block heater adjusted at 55 °C and capable of holding 13 × 100 mm glass tubes.

14. Sterile 13 × 100 mm borosilicate glass tube with caps (Fisher Scientific, Cat. No. 14-961-27).
15. Sterile salts stock solution for liquid inductions: 1 M $MgCl_2$ (Bioshop Canada, Cat. No. MAG510) + 1 M $CaCl_2$ (Bioshop Canada, Cat. No. CCL302).
16. Sterile salts stock solution for soft agarose: 3.8 M $MgCl_2$ (Bioshop Canada, Cat. No. MAG510) + 0.1 M $CaCl_2$ (Bioshop Canada, Cat. No. CCL302).
17. Sterile 96-well plates flat bottom (Sarstedt, Cat. No. 82.1581.001, or equivalent).

2.3 Phage Propagation in Broth and Soft Agarose

1. Pre-reduced TY liquid broth (3% tryptose, 2% yeast extract, pH 7.4).
2. Pre-reduced 1% TY bottom agar plates (TY with 10 g/L agar).
3. TY soft agarose (0.3 g/L agarose in TY) melted and kept at 55 °C in a block heater.
4. Sterile borosilicate glass tubes with caps suitable for optical density measurements (16 × 150 mm, Fisher Scientific, Cat. No.14-961-31).
5. Sterile borosilicate glass tubes with caps (13 × 100 mm, Fisher Scientific, Cat. No. 14-961-27).
6. Sterile 1.5-mL microcentrifuge tubes (Sarstedt, Cat. No. 72.690.001, or equivalent).
7. Sterile salt stock solution for liquid inductions: 1 M $MgCl_2$ (Bioshop Canada, Cat. No. MAG510) + 1 M $CaCl_2$ (Bioshop Canada, Cat. No. CCL302).
8. Sterile salt stock solution for soft agarose: 3.8 M $MgCl_2$ (Bioshop Canada, Cat. No. MAG510) + 0.1 M $CaCl_2$ (Bioshop Canada, Cat. No. CCL302).
9. Dry block heater adjusted at 55 °C and capable of holding 13 × 100 mm glass tubes.
10. Spectrophotometer set to a 600-nm wavelength (OD_{600}) with an adapter for glass tubes or a portable tube densitometer.
11. Sterile truncated 1000-μL pipet tips, or sterile Pasteur pipets.
12. Disposable 10-mL syringes (BD, Cat. No. 309604, or equivalent).
13. Disposable 0.22-μm or 0.45-μm syringe filters (Sarstedt, Cat. No. 83.1826.001, 83.1826, or equivalent).
14. Sterile 96-well plates flat bottom (Sarstedt, Cat. No. 82.1581.001, or equivalent).
15. Sterile 96-well plates V bottom (Sarstedt, Cat. No. 82.1583.001, or equivalent).

2.4 Phage Concentration with PEG and Genomic DNA Extraction

1. Conical 15-mL polypropylene centrifuge tubes (Sarstedt, Cat. No. 62.554.205, or equivalent).
2. Polyethylene glycol 8000 (BioShop Canada, Cat. No. PEG800).
3. NaCl 5 M stock solution (BioShop Canada, Cat. No. SOD004).
4. Disposable 10-mL serological pipettes (Sarstedt, Cat. No. 86.1254.001).
5. Sterile 2-mL microcentrifuge tubes (Sarstedt, Cat. No. 72.695.500, or equivalent).
6. Refrigerated centrifuge equipped with a microtube rotor capable of spinning at ≥20,000 × *g*.
7. Disposable 10-mL serological pipettes (Sarstedt, Cat. No. 86.1254.001).
8. Monarch® Genomic DNA Purification Kit (New England Biolabs, Cat. No. T3010S).
9. Agarose (BioShop Canada, Cat. No. AGA001).
10. Glacial acetic acid (Bioshop Canada, Cat. No. ACE222.4).
11. TAE 1 X buffer solution: 40 mM Tris Base, 20 mM acetate, and 1 mM EDTA.
12. Nanodrop DeNovix DS-11+.

2.5 Observation of Phage Particles by Transmission Electron Microscopy

1. Phage lysate with a minimum titer of 10^8 PFU/mL or more (ideally $>10^9$).
2. Disposable 10-mL syringes (BD, Cat. No. 309604, or equivalent).
3. Disposable 0.22-μm or 0.45-μm syringe filters (Sarstedt, Cat. No. 83.1826.001, 83.1826, or equivalent).
4. Ammonium acetate buffer stock solution, 1 M pH 7.0 (Bioshop Canada, Cat. No. ACE602). Sterilize on 0.22-μm filters. Dilute to 0.1 M working solution with sterile distilled water.
5. Sterile 1.5-mL microcentrifuge tubes (Sarstedt, Cat. No. 72.690.001, or equivalent).
6. Refrigerated centrifuge equipped with a microtube rotor capable of spinning at ≥20,000 *g*.
7. Uranyl acetate 2% solution. Filter-sterilize on min μm membrane. Store at 4 °C for a maximum of 1 week.
8. Formvar–/carbon-supported copper grids, 400 mesh (Sigma-Aldrich, Cat. No. 931047).

3 Methods

3.1 In Silico Prophage Identification

Several bioinformatic tools have been developed for the prediction of prophages in bacterial genomes. Some of these include PHASTER, PhiSpy, Phigaro, VIBRANT, PropheET, and VirSorter2. The strategies and prediction algorithms vary between these tools, as well as the type of input data. For example, PHASTER offers a user-friendly web interface (phaster.ca) that makes it easy to analyze a single or a few bacterial genomes. For larger datasets, standalone tools that can be run locally on a personal computer or on computer clusters are preferable. These tools often require some computer skills for installation and running on Linux or Mac OS, as they do not offer graphic user interface and must be run in command line. In our lab, we often use PHASTER or PHASTEST (an improved version of PHASTER), when only one or a few genomes must be analyzed (https://phaster.ca; https://phastest.ca). For those familiar with command line, a Docker image can also be downloaded to run PHASTEST locally. When we need to analyze several bacterial genomes, we run geNomad (https://github.com/apcamargo/genomad/), a powerful tool capable of dealing with thousands of bacterial genomes to identify plasmids and prophages. In a benchmark study, geNomad outperformed other common prophage identification software [4]. The tool is available as a web application on the Galaxy and NMDC EDGE platforms, where sequences can be uploaded and analyzed. geNomad can also be installed and run locally on a personal computer. For larger datasets, installation on computer clusters is possible and preferable.

The output files of PHASTER (or PHASTEST) will indicate whether the identified prophage is intact (score > 90), probable (score 70–90), or incomplete (score < 70). A genome viewer can help visualize the prophage in the bacterial genomic context. Several output files are generated by geNomad, including summary.tsv files that list all plasmids and all proviruses detected by geNomad. The tables indicate the provirus coordinates in the bacterial genome or contig, the score (between 0 and 1, a higher score being better), and the number of plasmid and prophage hallmark genes. geNomad will also automatically extract the prophage sequence in fasta format (both DNA and proteins), which streamlines downstream analyses.

Note that in *C. difficile*, diffocins are frequently present in genomes and their size is around 20 kb. Although these clusters are indicated as incomplete in PHASTER analyses, diffocins are nonetheless active and can produce phage tail-like bacteriocins [5]. It is important to manually review the predicted prophages and to look for characteristic gene products. Functional *C. difficile* phages normally carry genes encoding terminase, capsid, and tail functions, in addition to regulatory elements. If capsid or tail genes are absent from the prophage sequence, it could indicate a non-functional prophage or a diffocin.

3.2 Prophage Induction in Liquid Broth

3.2.1 Mitomycin C or Antibiotic Induction

DNA-damaging agents, such as mitomycin C or antibiotics like fluoroquinolones, are commonly used to induce prophages. As all known *C. difficile* phages are temperate, prophage induction is a staple method for phage isolation and characterization. However, not all prophages will respond to inducing agents or to the same treatment, and sometimes different strategies need to be tested, such as UV induction (next section). Examples of this are the prophages of strain CD24. Indeed, phiCD24-1 can only be induced by mitomycin C while UV induction must be used for phiCD24-2 [6, 7]. Below, we describe a classic method for prophage induction with mitomycin C in broth, with optical density monitoring.

1. Inoculate 5 mL of pre-reduced TY broth, in 16 × 150 mm glass tubes, with a fresh isolated colony or a frozen glycerol stock (−80°C). Incubate overnight (O/N) at 37 °C under anaerobic conditions (*see* **Note 1**).
2. The next day, inoculate 4 × 10 mL TY broth with 1% of the O/N culture.
3. Incubate at 37 °C and monitor the optical density at 600 nm (OD_{600}) using a spectrophotometer inside the anaerobic chamber.
4. When the cultures reach OD_{600} = 0.1 (early exponential growth phase), add different subinhibitory concentrations of mitomycin C or another desired DNA-damaging antibiotic (*see* **Note 2**) to three of the four tubes, leaving the fourth as a growth control.
5. Incubate at 37 °C and continue monitoring OD_{600} regularly (at 30–60 min intervals). During the incubation process, OD_{600} should rise steadily for 2–3 h after which a sharp drop will be observed due to phage induction-mediated lysis (*see* **Note 2**). Proceed to the next steps once the OD_{600} stabilizes after the drop.
6. Take the tubes out of the anaerobic chamber and transfer cultures into 15-mL conical centrifuge tubes and spin at ≥3000 × *g* for 10 min at 4 °C. Centrifugation will pellet intact cells and debris.
7. Filter the supernatant through a 0.22-μm or 0.45-μm syringe filter to obtain an induction lysate containing phage particles (*see* **Note 3**). Store the lysate at 4 °C until use or proceed directly to the identification of a replicating host (Subheading 3.3).

3.2.2 Prophage Induction in Liquid Broth Using Ultraviolet Light

Some prophages will be unaffected by mitomycin C or other DNA-damaging antibiotics such as fluoroquinolones [8, 9]. Another method for prophage induction uses ultraviolet light (UV). The procedure below describes UV prophage induction with OD_{600} monitoring.

1. Prepare four tubes of bacterial culture for induction as specified in **steps 1–3** of Subheading 3.2.1.
2. As the OD_{600} reaches 0.2, anaerobically transfer each of the cultures into 15-mL conical tubes and screw caps tightly before removing from the anaerobic chamber.
3. Centrifuge at ≥3000 × *g* for 10 min in a refrigerated centrifuge at 4 °C.
4. Inside the chamber, discard the supernatant and resuspend the bacterial pellets in 100 μL of pre-reduced TY broth.
5. Use sterile 3-mm or 6-mm glass beads or a cell spreader to uniformly spread the 100 μL onto the surface of a pre-reduced TY agar plate. Allow the liquid to fully absorb before proceeding to the next step.
6. Outside the anaerobic chamber, place the plates upside down and without the lid on a UV transilluminator. Irradiate three of the four plates at a wavelength of 302–365 nm for 5, 10, and 15 s, respectively. Do not irradiate the fourth plate, which serves as a positive control for bacterial growth. Once the irradiations are done, immediately cover the plates and quickly return them inside the anaerobic chamber. Avoid prolonged exposure to oxygen.
7. Flood the surface of the irradiated plates with 5 mL of pre-reduced TY broth to collect bacterial cells. Transfer the cell suspension into 16 × 150 mm glass tubes.
8. Follow **steps 5–7** from Subheading 3.2.1.

3.3 Identification of a Susceptible Host for Phage Propagation

To identify bacterial strains susceptible to infection and suitable for phage propagation, we use the soft agar overlay method in spot assays or full plaque assay.

3.3.1 Soft Agar Overlay

1. Inoculate 5 mL of pre-reduced TY broth, in 16 × 150 mm glass tubes, with a fresh isolated colony or a frozen glycerol stock (−80°C). Incubate O/N at 37 °C under anaerobic conditions (*see* **Note 1**).
2. The next day, dilute the culture of the phage host strain (1–3%) into 5 or 10 mL of fresh pre-reduced TY broth. Incubate anaerobically at 37 °C until the OD_{600} reaches 0.3–0.6, corresponding to mid-exponential phase (*see* **Note 4**).
3. Meanwhile, melt TY soft agar using a microwave, loosening the bottle cap slightly to prevent pressure buildup. Dispense 3 mL

of the melted soft agar into 13 × 100 mm sterile glass tubes. As melting removes oxygen, no additional medium pre-reduction is needed at this stage. Keep the tubes at 55 °C in a dry block heater inside the anaerobic chamber until use. To minimize evaporation, avoid preparing soft agar too far in advance.

4. To each soft agar tube add 0.1 mL of a 3.8 M $MgCl_2$ + 0.1 M $CaCl_2$ solution and 0.5 mL of the log-phase (OD_{600} = 0.3–0.6) culture to be tested (*see* **Note 5**). Mix gently and then quickly pour on top of a pre-reduced TY agar plate avoiding bubbles. Gently rock the plate to ensure a uniform spread of the soft agar overlay. Place the plate on a level surface and allow the agar to solidify.
5. While the overlays set, prepare serial ten-fold dilutions of phage lysates by adding 20 μL of phage lysate to 180 μL of TY broth using a 96-well plate. For the spot test assay, follow **step 6**. Otherwise, for the phage plaque assay follow **step 7**.
6. *Spot test assay.* On a level surface, pipette 5-μL drops of each phage lysate dilution side by side on top of the solidified soft agar. Ensure sufficient spacing to prevent droplet merging. Leave plates on a level surface until droplets are fully absorbed. Once dry, incubate the plates upside down O/N at 37 °C under anaerobic conditions (*see* **Note 6**) (Fig. 1).
7. *Full plaque assay.* While the spot test is useful for screening multiple phage lysates on a single strain, it gives only a rough estimation of phage sensitivity or phage titer. The full plaque assay enables better observation and counting of individual plaques. Additionally, plaque size can provide insight into phage fitness on a particular strain. For instance, larger plaques

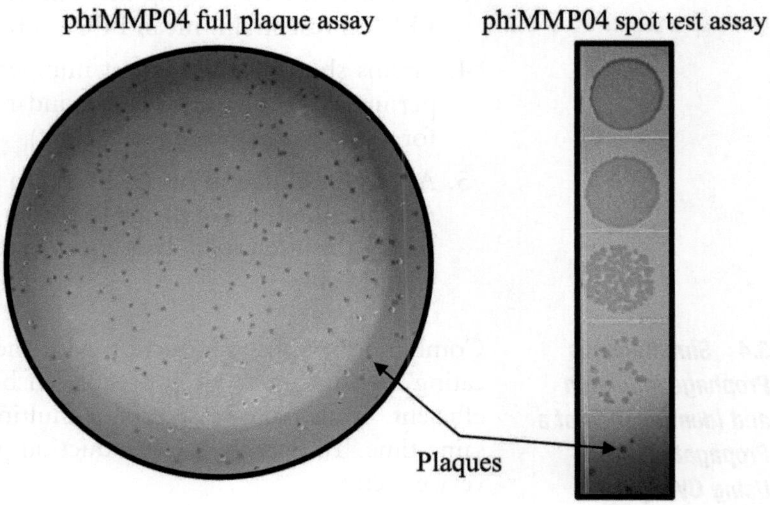

Fig. 1 Phage phiMMP04 plaques obtained from full plaque and spot test assays on a susceptible host

may indicate enhanced phage performance on the tested strain. To perform the full plaque assay, add 100 μL of the desired phage dilution directly to the soft agar tube at **step 4** before pouring onto the plate. Testing multiple phage dilutions is critical to identify plates with well-isolated, countable plaques (ideally 30–300 PFU) (Fig. 1).

8. After an O/N incubation, bacterial strain susceptibility to the phage lysate can be ascertained by observation of lysis zones for the spot test assay (*see* **Note 6**), or the appearance of isolated plaques in the soft agar for the full plaque assay. For smaller plaques, use of a colony counter with a magnifying lens to help visualization.

3.3.2 Determination of the Efficiency of Plaquing (EOP)

When screening for a susceptible host, some strains might be infected, but multiple factors can lead to poor phage propagation. Spot test assays with concentrated phage lysates might not reveal this clearly, so determining the efficiency of plaquing (EOP) can be useful. This is done following the full plaque assay procedure described in Subheading 3.3.1, **step** 7.

1. Following incorporation of serial ten-fold dilutions of phage lysates into soft agar overlays, choose the dilutions yielding well-isolated plaques, typically 30–300 PFU per plate.
2. For each phage tested, the number of PFU is counted on both a reference strain, which can be the strain initially used for phage isolation or any other strain, and the test strain to compare with. Perform all EOP measurements in biological triplicates, each in technical triplicates, and report the results as mean EOP values ± standard deviation.
3. Calculate the EOP using the following formula: EOP = mean PFU on test strain/mean PFU on reference strain.
4. Strains showing the highest number of plaques are the most permissive to phage infection and represent good candidates for a propagating host (see below).
5. A rough estimation of the EOP can also be determined using multiple dilutions of phage lysates in spot test assays and comparing the intensity of lysis zones, or counting isolated plaques, when visible.

3.4 Simultaneous Prophage Induction and Identification of a Propagating Host Using UV Light

Combining prophage induction with the identification of a replicating host into a single protocol can be advantageous and time efficient. It also allows screening multiple lysogenic hosts at the same time. To do this, a UV induction procedure on agar plate is very effective.

1. In a 16 × 100 mm glass tube, inoculate 5 mL of pre-reduced TY liquid broth using either a fresh isolated colony or frozen

glycerol stock. Inoculate one tube for each of the strains to be tested either for induction or phage susceptibility. Incubate in the anaerobic chamber at 37°C O/N (*see* **Note 1**).

2. The next day, prepare 3× 900 mL aliquots of pre-reduced TY liquid broth in microcentrifuge tubes for each strain to be tested for prophage induction. Using the O/N cultures, prepare ten-fold serial dilutions of each strain up to 10^{-3}. Dilutions should be handled in the anaerobic chamber.
3. Drop 5 μL of each of the dilutions (10^{-1}, 10^{-2}, and 10^{-3}) for each strain to induce on top of pre-reduced TY bottom agar plate. At least 4 to 5 strains can be arrayed side-by-side on a single plate. Incubate at 37 °C in the anaerobic chamber for 4 h to let cells grow to exponential phase (*see* **Note 7**).
4. Meanwhile, transfer 0.5 mL of the potential propagating host strains in 5 mL of pre-reduced TY broth and incubate at 37 °C under anaerobic conditions. After 4 h, the cultures should have reached early log phase ($OD_{600} = 0.3–0.6$) (*see* **Note 4**).
5. Prepare melted soft agar as in **step 3** of Subheading 3.3.1.
6. After the spotted plates have been incubated for 4 h, remove the plates from the anaerobic chamber and lay them face down without the lid on a UV transilluminator.
7. Irradiate the plates for 10 s at 302–365 nm and quickly bring them back inside the anaerobic chamber. Different irradiation times can be tested to find the best conditions.
8. Combine the $MgCl_2$ and $CaCl_2$ mix (*see* **Note 5**) and the potential host strain with the melted soft agar and then pour on top of the irradiated plates as described in **step 4** of Subheading 3.3.1. However, be careful not to rock the plates too vigorously when spreading the soft agar overlay, to avoid mixing the bacterial spots. Allow the soft agar to harden and dry on a level surface for 5–10 min.
9. Incubate at 37 °C in anaerobic conditions for 24 h.
10. If the spotted strains carry inducible prophages that can infect the strain within the soft agar, lysis zones or plaques will appear in the soft agar overlay above the corresponding spots. Plaques or lysis zones can be harvested with a truncated 1-mL pipet tip or a Pasteur pipet. Phage can then be purified following the method below in Subheading 3.5.

3.5 Phage Purification by Plaque Assay

Once a host strain has been identified for a phage, it is necessary to purify the phage for downstream applications, and three successive rounds or purification are generally recommended.

1. Inoculate 5 mL of pre-reduced TY liquid broth with a fresh colony or a frozen glycerol stock of the phage propagating host strain and incubate O/N at 37 °C in anaerobic conditions (*see* **Note 1**).

2. The next day, use a 1-mL truncated pipet tip or a Pasteur pipet to transfer an isolated lysis plaque or part of a lysis zone in the form of an agar plug into 0.5 mL of TY broth in a microcentrifuge tube. Leave at room temperature for at least 3 h for phage diffusion or O/N at 4°C.
3. Centrifuge the microtube at 13,000 × *g* for 1 min to pellet bacterial cells and debris and transfer the supernatant to a new microcentrifuge tube. Alternatively, filter the phage diffusion using a 0.22-μm or 0.45-μM syringe filter. Before filtering, wet the filter with sterile phage buffer to avoid losing phages in the void volume. Store the supernatant or filtered fraction at 4°C until use.
4. Inoculate a 5-mL culture in TY broth using 1–3% of the O/N culture and incubate at 37 °C under anaerobic condition until the OD_{600} reaches 0.3–0.6 (*see* **Note 4**).
5. Meanwhile, prepare 0.3% soft agar overlays and ten-fold serial phage dilutions (up to 10^{-4}) as described in Subheading 3.3.1 using the plaque assay method. Combine the log phase culture of the propagation strain and the phage dilutions (one dilution per tube). Once the overlay has been poured and fully solidified, incubate in anaerobic conditions for 24 h at 37 °C.
6. The next day, repeat the above procedure two more times using single isolated phage plaques each time. The last plate should contain a pure phage.

3.6 Phage Propagation: Primary Amplification from Purified Phage Plaques

Most phage experiments require high titers (e.g., $\geq 10^9$ PFU/mL). In addition, concentrated phage lysates are more stable than diluted ones. Therefore, it is important to amplify the isolated phages. With *C. difficile* phages, a broth amplification protocol works well, and titers $>10^9$ and up to ~10^{10} PFU/mL can be routinely obtained.

1. Inoculate 5 mL of TY broth using a fresh isolated colony or a frozen glycerol stock (−80 °C) of the sensitive host. Incubate O/N under anaerobic atmosphere at 37 °C (*see* **Note 1**).
2. Prepare 5 or 10 mL of TY broth in a glass tube. Add one or a few phage plugs from the last phage purification plate (Subheading 3.5; *see* **Note 8**) and supplement with $MgCl_2$ and $CaCl_2$ to a final concentration of 10 mM each (*see* **Note 5**). Inoculate with 1% of the O/N culture of the appropriate *C. difficile* host strain and incubate O/N at 37 °C under anaerobic conditions. In the best-case scenario, the cultures will be cleared, but it is more common to observe incomplete lysis or bacterial regrowth, and therefore, turbid cultures are often obtained.
3. The following day, transfer the culture into a 15-mL conical tube and centrifuge at 4000 × *g* for 10 min to pellet intact bacteria and debris.

4. Collect the supernatant and filter it through a 0.22-μm or 0.45-μm syringe filter. The filtered lysate can be stored at 4 °C until further use or frozen at −80 °C in 50% glycerol for long-term storage [10].
5. Titration of the primary amplification can be done by adding serial ten-fold dilutions of the phage lysate into soft agar overlays and counting the number of phage plaques, as described in Subheading 3.3.2. Alternatively, the spot test assay can be used to get a rough estimation of the phage titer.

3.7 Phage Propagation: Secondary Amplification

Following a primary amplification, phage titers are typically in the range of 10^6 to 10^8 PFU/mL. Therefore, a second amplification step using the primary amplification lysate as a "seed" is recommended to increase the phage yield. However, if the phage titer already exceeds 10^9 PFU/mL, a second propagation step is not required. Optimization of the multiplicity of infection (MOI, or the phage-to-bacteria ratio), is important to maximize phage yield (*see* **Note 9**). Also, to minimize the risk of genetic drift, a good practice is to perform all secondary amplifications using the same primary amplification stock and not using consecutive amplifications as the phage seed.

1. Inoculate 5 mL of TY broth with a freshly isolated colony or a glycerol stock of the sensitive host (stored at −80 °C). Incubate O/N at 37 °C under anaerobic conditions (*see* **Note 1**).
2. The following day, inoculate 4 glass tubes containing 10 mL TY broth with 1 or 3% of the O/N culture.
3. Monitor the optical density (OD_{600}) using a spectrophotometer under anaerobic conditions.
4. Once the OD_{600} reaches 0.1 (*see* **Note 2** regarding the optimal OD), supplement each tube with $MgCl_2$ and $CaCl_2$ to a final concentration of 10 mM each (*see* **Note 5**). Then, inoculate three of the four tubes with 1, 0.1, and 0.01 mL of the first amplification phage lysate. The fourth tube will not contain phages and will serve as a positive control for bacterial growth.
5. Monitor the OD_{600} at regular intervals (e.g., every 30–60 min). The culture density should increase for about 3 h or more and then sharply decline as cell lysis occurs. A marked drop in OD_{600} generally indicates good phage amplification. Wait until the OD_{600} stops decreasing for two consecutive readings and then proceed to the next step. Avoid incubating the infected cultures for longer periods, as bacterial regrowth often occurs. This indicates phage entry into the lysogenic cycle, which could reduce the final titer. Amplification typically completes within 5 to 8 h.
6. Transfer the lysate into a 15-mL conical tube and centrifuge at 4000 × *g* for 10 min at 4 °C to pellet intact bacteria and debris.

7. Collect the supernatant and filter it using a 0.22-μm or 0.45-μm syringe filter and store at 4 °C until use or proceed with phage titration as described in Subheading 3.3.2.

3.8 Phage Adsorption Assay

The adsorption assay is a very useful test because it provides important information about phage–host interactions [11]. For example, adsorption can be used to determine if a host strain is resistant to phage infection because it does not express the appropriate phage receptor (hence no phage binding), or if other mechanisms interfere with phage propagation, like antiphage systems. It is also highly useful when investigating bacteriophage-insensitive mutants (BIMs).

1. Inoculate a 5-mL preculture of the *C. difficile* strain to be tested for phage adsorption in TY broth and incubate O/N at 37 °C under anaerobic conditions.
2. In parallel, inoculate another 5-mL tube containing TY broth with the *C. difficile* phage host strain (used for titration in soft agar overlay) and incubate O/N at 37 °C anaerobically.
3. The next day, dilute the O/N culture of the phage host strain (1 or 3%) into 5 or 10 mL of fresh TY broth and incubate anaerobically at 37 °C until the OD_{600} reaches 0.3–0.6 (*see* **Note 3**).
4. Meanwhile, prepare the phage lysate by diluting the phage stock to a final concentration of 10^5 PFU/mL in TY medium. To do this, perform serial ten-fold dilutions by transferring 100 μL of the phage suspension into 900 μL of TY medium, ensuring thorough mixing by inverting the tube 20 times at each step to achieve proper homogenization (*see* **Note 10**). This step can be performed under aerobic conditions.
5. Inside the anaerobic chamber, prepare a negative adsorption control by combining in a microtube containing TY medium $MgCl_2$ and $CaCl_2$ to a final concentration of 10 μM each. Then add 100 μL of phage lysate (10^5 PFU/mL). The final volume should be 1 mL.
6. Next, prepare technical replicates by combining 900 μL of the O/N culture of the *C. difficile* test strain with 100 μL of phage lysate (10^5 PFU/mL) in separate tubes (minimum three replicates, under anaerobic conditions).
7. Mix all tubes (one negative control + three technical replicates) thoroughly by inverting 20 times. Incubate the tubes at 37 °C in the anaerobic chamber for 30 min (*see* **Note 11**).
8. Meanwhile, prepare soft agar overlays following the standard phage titration protocol (Subheading 3.3.1).
9. After incubation, remove the tubes from the incubator and centrifuge at 13,000 × *g* for 1 min. Transfer 100 μL of the

supernatant from each tube to a fresh microtube and discard the remaining pellet. Bring the supernatants back inside the anaerobic chamber and add 900 μL of reduced TY broth and mix thoroughly by inverting 20 times. This step results in a tenfold dilution of the supernatant prior to titration.

10. Titrate the free phages that did not adsorb using the phage host strain in full plaque assay, as described in Subheading 3.3.1. Plate each technical replicate and negative control in triplicate. Include a titration control plate prepared with the phage host strain only (no phage), to confirm proper bacterial lawn formation (*see* **Note 12**).
11. Incubate all plates O/N at 37 °C under anaerobic condition.
12. The following day, count the number of plaques on each plate. Calculate the percentage of adsorption using the formula: 100—[(residual titer/initial titer) × 100].
13. Repeat the entire experiment independently three times. Present the data as mean values ± standard deviation.
14. As an alternative to the above protocol, a high-throughput adsorption assay can be carried out in a V-bottom 96-well plate to screen multiple phage-bacteria combinations simultaneously.
    (a) In each well, combine 180 μL of the test bacterial culture supplemented with $MgCl_2$ and $CaCl_2$ (final concentration: 10 μM each) and 20 μL of phage lysate at 10^4 PFU/mL, for a total volume of 200 μL.
    (b) Include three technical replicates per condition and a negative control containing 180 μL of TY broth with salts instead of the bacterial culture.
    (c) Incubate the plate at 37 °C for 30 min in the anaerobic chamber.
    (d) After incubation, centrifuge the plate at 2200 rpm for 3 min outside the anaerobic chamber.
    (e) Using a multichannel pipette, transfer 100 μL of supernatant from each well into a new 96-well plate.
    (f) Proceed with a spot test assay as described in Subheading 3.3.1 to assess the presence of residual phages. Note that since adsorption tests are performed with smaller volumes, more variability can be expected in the results.

3.9 PEG Precipitation and Genomic DNA Extraction of Viral Particles for Whole Genome Analyses

Extraction of total phage DNA is necessary to further characterize the isolated phages using restriction enzymes or whole phage genome sequencing. We generally perform a PEG precipitation of viral particles to concentrate them before extraction to increase the overall DNA yield. Phage genomic DNA can then be extracted by different ways, but we recommend using a commercial kit that provides highly pure DNA. In our lab, we use the Monarch® Genomic DNA Purification Kit from New England Biolabs.

1. For PEG precipitation, transfer 5 mL of a high-titer phage lysate (≥10^8 PFU/mL and ideally ≥10^9 PFU/mL) into a 15-mL centrifuge tube.
2. Add 1 mL of 5 M NaCl and 0.6 of polyethylene glycol 8000 (PEG 8000) and mix by inversion until the PEG is completely dissolved.
3. Divide the phage lysate into 3× 2 mL microcentrifuge tubes and incubate at 4 °C for a minimum of 3 h, but O/N incubation is recommended.
4. The next day, centrifuge the tubes at 14,000 × *g* for 30 min at 4 °C. Carefully discard the supernatant and resuspend every phage pellet in a single 0.2-mL volume of phage buffer.
5. Proceed with DNA extraction using the Monarch® Genomic DNA Purification Kit, following the protocol for Gram-positive bacteria starting at **step 4** (incubation with proteinase K). Elute the DNA in 25 μL of elution buffer (e.g., 10 mM Tris–HCl pH 8.0).
6. Check the quality and concentration of the purified DNA by Nanodrop or Qubit and by running an aliquot on a 0.8% agarose gel to monitor any potential degradation. Starting from a 5 mL lysate at ~10^9 PFU/mL, you should expect at least a few μg of pure DNA at a concentration of at least ~100 ng/μL.

3.10 Analysis of Phage Genomes with Restriction Enzyme

Although sequencing technologies have become more accessible over the years, a quick and low-cost restriction enzyme analysis is often useful to confirm the identity of a known phage, or to discriminate phage isolates.

1. In a microcentrifuge tube, use the appropriate restriction endonuclease to digest 1 μg of phage genomic DNA following the manufacturer's recommendations. We routinely use *Hind*III as this enzyme generates a reasonable number of bands.
2. After digestion, incubate the DNA at 75 °C for 10 min to dissociate potential cohesive termini within the phage genome. Rapidly chill on ice (*see* **Note 13**).
3. Run the whole digest on a 0.8% agarose gel at a constant voltage (110 V) in 1X TAE buffer. Post staining of the gel is recommended. Reveal profiles using a gel documentation system.

3.11 Whole Phage Genome Sequencing

Whole genome sequencing can be performed with both short-read (e.g., Illumina or Aviti) and long-read technologies (e.g., Oxford Nanopore, PacBio). The advantage of using short-read sequencing is that bioinformatic tools such as PhageTerm [12] or PhageTerm-Virome [13] can be used to determine the type of phage genomic termini from the sequence data (e.g., *cos*, *pac*, direct terminal repeat). To be able to use these tools however, the sequencing library must be prepared using randomly fragmented DNA (e.g., Covaris, sonication) or library preparation kits that preserve the natural DNA ends (e.g., Illumina TruSeq PCR free, TruSeq Nano). Sequences generated with ONT or PacBio can in principle be analyzed with PhageTermVirome, as long as the natural DNA ends are preserved during library synthesis (e.g., Native barcoding kit). In our lab, we have used both strategies with success, and we are now using Oxford Nanopore on a routine basis.

3.12 Bioinformatic Analyses of Whole Phage Genomes

Multiple tools have been developed to annotate, compare, and analyze phage genomes. Here we present the pipeline that we routinely use to analyze our phage genomes. First, we use Pharokka [14] to identify coding sequences and perform a first annotation of the genome. Pharokka is based on Prokka [15], developed for bacterial genome annotation, and it uses PHANOTATE [16], a phage-optimized tool to predict coding sequences in phage genomes. Pharokka also interrogates the PHROG database for functional annotation [17] and searches for tRNAs, tmRNAs, and CRISPRs (Clustered Regularly Interspaced Short Palindromic Repeat sequences) using tRNAscan-SE 2 [18], Aragorn [19], and MinCED [20], respectively. It also searches for virulence factors and antimicrobial resistance genes through interrogation of the Virulence Factor DataBase (VFDB) [21] and the Comprehensive Antibiotic Resistance Database (CARD) [22]. Various output files are generated by Pharokka, including a list of the identified CDS and their predicted functions and any virulence factor or antibiotic resistance gene. Fasta and Genbank files of the genome are also produced and the sequence can be reorganized so that the genome starts with the terminase gene, therefore facilitating multiple genome comparison.

Phage genomes often encode multiple genes for which no function can be predicted with current annotation tools (including Pharokka). PHOLD was recently developed to improve phage annotation and solve this problem [23]. This tool uses the ProstT5 protein language model [24] and Foldseek [25] to compare the predicted structure of phage proteins to a curated database of 1.36 million protein structures, mostly derived from phages. PHOLD can be run locally with command line, and it is recommended to have a dedicated graphic card to use the computer's GPU, which greatly speeds up the analysis to less than a minute for a 50-kb phage genome. Recently, another phage annotation tool was developed: Empathi [26]. We have not tested this tool yet, but based on the benchmarks and published data, it outperforms PHOLD and all the other annotation tools and can greatly improve phage annotation.

To compare multiple phage genomes, we use vConTACT v.2.0 [27] and Cytoscape [28] to cluster and visualize phages, based on their level of shared proteins. This allows regrouping closely related phage genomes into clusters and to display them within a more complex network. Specific clusters can then be aligned and visualized using clinker [29, 30].

3.13 Observation of Phage Particles Under the Electron Microscope

With recent revisions made by the International Committee on Taxonomy of Viruses (ICTV), phage morphology is no longer a criterion for phage classification. For instance, the *Siphoviridae*, *Myoviridae*, and *Podoviridae* families have been abandoned, and genomic data is used for phage classification. Yet, observation of phage particles by transmission electron microscopy (TEM) can still be relevant to identify phage structures and appendages that cannot be readily identified using bioinformatic tools. Using a concentrated phage lysate ($\geq 10^9$–10^{10} PFU/mL) purified by CsCl is recommended to obtain high-quality TEM pictures. However, if multiple phage preparations need to be observed, a rapid method can be more advantageous. Below is a simple and effective protocol to quickly visualize phage particles by TEM, without the need for complex phage purification.

1. Transfer 1.5 mL of a concentrated phage lysate ($\geq 10^9$ PFU/mL) into a sterile microcentrifuge tube.
2. Spin at ~20,000 × *g* for 1 h, at 4 °C.
3. Gently pipet and discard the supernatant, leaving 100 μL at the bottom.
4. Add 1.5 mL of ice-cold ammonium acetate working solution (0.1 M, pH 7.0).
5. Spin at ~20,000 × *g* for 1 h, at 4 °C.
6. Repeat **steps 3–5** another time.
7. Discard the supernatant and keep the bottom 100 μL. Store at 4 °C until use or directly prepare TEM grids (*see* **Note 14**).
8. For staining, different options can be tested, and conditions might need to be adjusted depending on your material. Here are two variations that we routinely use and that generally give satisfactory results (*see* **Note 15**).
9. *Option 1*: Using clean precision tweezers, place 5 μL of a drop of washed phages on the matte side of a TEM grid. Mix in 5 μL of freshly prepared 2% uranyl acetate (UA) by pipetting up and down a few times. Let stain for 1 min and then blot with a Whatman paper by touching the side of the grid. Let dry a few minutes and observe.

10. *Option 2*: Place a drop of phage preparation on the mate side of a grid and let bind for 1 min. Blot with a Whatman paper by touching the side of the grid. Add 5 μL of 2% UA solution and let stain for 1 to 5 min (adjust the staining by trial and error). Blot the liquid as above and then add a 10 μL drop of ultrapure water and let stand for 1 min. Blot the liquid as above and let dry for a few minutes before observation.

3.14 Creation of New Lysogens

The quest for strictly lytic phages against *C. difficile* has always been a topic of great interest. Relying on the simple observation of clear phage plaques on agar plates does not guarantee that the phage is strictly lytic. Some *C. difficile* phages tend to form clear plaques on agar, and still, they are true temperate phages. Whole genome sequencing can generally help determine if a phage is strictly lytic or temperate. For instance, the absence of lysogeny-associated genes such as integrase, CI, or Cro repressors can suggest that the phage is strictly lytic. Yet, experimental confirmation that lysogeny does not occur during infection is recommended. In addition, when studying new prophages, it can be interesting to determine their role in the biology of their host. In fact, previous reports have shown that certain *C. difficile* phages can alter toxin production, sporulation, and biofilm formation [31–33]. Below is a simple protocol to create lysogens. It relies on the principle that a high MOI generally promotes lysogeny. Our experience has shown that using this protocol, a very high frequency of lysogeny can be reached with certain phage–host pairs (sometimes close to 100%).

1. The day before, inoculate 5 mL of TY broth in a 16 × 150 mm glass tube with either fresh isolated colonies or a frozen glycerol stock (−80 °C) with the desired test strains. Incubate at 37 °C O/N in anaerobic conditions (*see* **Note 1**).
2. The next day, inoculate a fresh 5-mL tube of TY broth with 3% of the O/N culture. Incubate at 37 °C until the OD_{600} reaches 0.3–0.6.
3. Prepare soft agar overlays as described in Subheading 3.3.1. However, replace the bacteria in the soft agar with 100 μL of a high-titer phage lysate (≥10^9 PFU/mL) corresponding to the phage for which you want to assess its lysogenic potential. Prepare two or three of these plates in addition to an equal number of plates without the phage. Let dry on a leveled surface for 5–10 min.
4. Prepare microcentrifuge tubes with 0.5 mL of sterile TY broth. Prepare serial ten-fold dilutions of the mid-log phase cultures to be tested up to 10^{-5}. To proceed to the next step, choose dilutions that should form ~100–1000 CFU per plate.
5. Spread 0.1 mL of the chosen dilutions in duplicates (each on a plate with and without phage) using sterile 3-mm glass beads. During spreading, rock the plates gently to avoid shearing the soft agar.

6. Remove the beads and incubate upside down for 24–48 h at 37 °C in the anaerobic chamber.
7. The next day, compare the colony counts between phage containing plates and plates without phage. Depending on the phage's propensity for lysogeny, dozens to hundreds of lysogens can be expected.
8. Pick two or three colonies and streak them out on TY agar bottom plates. Incubate at 37 °C O/N. Purify these colonies by streaking them again for two more rounds using the same method. This will ensure that no residual phage particles are present on the final plates.
9. Confirmation of lysogeny can be done by inducing the prophage with the appropriate inducer as in Subheading 3.2, by showing that the clones are resistant to the original phage (a hall mark of lysogeny), and/or by amplifying a region of the phage genome using specific PCR primers. If the phage is known to integrate into the bacterial chromosome and if the integration site is known, choosing primers that overlap the bacteria–phage junction is best to confirm integration, but this will not work with episomal prophages.

4 Notes

1. Being an obligatory anaerobe, *C. difficile* is highly oxygen-sensitive, and prolonged exposure to atmospheric conditions can severely impact its viability. To promote optimal growth, it is strongly advised to use pre-reduced media. This can be achieved by placing agar plates and unscrewed bottles or tubes with liquid medium inside the anaerobic chamber for at least 3–4 h, and ideally O/N before use. Based on our experience, using a gas mixture composed of 10% H_2, 5% CO_2, and 85% N_2 provides more stable and efficient anaerobic conditions. However, institutional safety policies may limit the allowable hydrogen concentration in anaerobic systems to 5%.
2. Many DNA-damaging agents have been used for prophage induction in *C. difficile*. It has been shown that prophages of this bacterium can be induced by mitomycin C, norfloxacin, ciprofloxacin, moxifloxacin, and levofloxacin. We use up to five different concentrations of mitomycin C for optimal induction. Alternatively, it is possible to determine the appropriate concentration range of an inducer by determination of its minimum inhibitory concentration (MIC). Aim for a concentration the does not cause to much growth inhibition, but high enough to induce the prophage. Final concentrations of 0.5–3 μg/mL of mitomycin C and norfloxacin are generally

optimal for prophage induction. If a strain is less susceptible to the inducer, higher concentrations might be required. Using the right concentrations of inducers is key for a good phage induction. The timing of the addition of the inducing agent is also important. We recommend adding the inducer when cultures reach an OD_{600} of 0.1 or maximum 0.15. Inducing at higher ODs will often lead to poor induction. Aim for conditions where cultures will grow up to an OD_{600} of 0.2–0.6 before dropping by around 0.2–0.3 OD_{600} units. Such conditions should yield a productive induction. After the visible drop, the OD_{600} will most likely not reach 0. Rather, a plateau or culture rebound is often observed instead. As such, when the OD_{600} has stabilized for at least 30–60 min, we recommend centrifuging and filtering the lysate rather than leaving the tubes to incubate O/N.

3. Both 0.22-μm and 0.45-μm membranes can be used to filter *C. difficile* phage lysates. Although 0.22-μm filters are adequate for most phages, a 0.45-μm pore size can facilitate filtration of viscous or high-titer lysates by reducing filter clogging.
4. The *C. difficile* phage host strain (or infection test strain) should be harvested at mid-exponential phase, typically at an $OD_{600} = 0.3–0.6$. Cultures at this density provide the best results, yielding smooth and uniform bacterial lawns when overlaid in soft agar. This ensures clear and countable plaques for accurate titration. Using stationary-phase cultures or outside this range can result in sparse lawns, uneven or overly thick growth, reduced phage diffusion, and smaller plaques, compromising titration accuracy. Also, it is highly recommended to pour all soft agar overlays under anaerobic conditions, minimizing oxygen exposure and ensuring uniform *C. difficile* growth.
5. Cationic ions such as Mg^{2+} and Ca^{2+} are often essential for promoting phage adsorption and successful infection [34, 35]. In the case of *C. difficile* phages, the presence of high $MgCl_2$ concentrations in soft agar overlays is crucial for certain phages [36]. For example, ϕCD38–2 only forms plaques when 0.4 M $MgCl_2$ is present [32]. While not all phages require such elevated salt levels, we routinely use a combination of 0.4 M $MgCl_2$ and 10 mM $CaCl_2$ in TY soft agar to ensure consistent propagation of all our *C. difficile* phages. For amplifications carried out in TY broth, both salts are also necessary, though at lower concentrations (10 mM each of $CaCl_2$ and $MgCl_2$).
6. When performing and interpreting results from spot test assays, we must be aware of the "lysis from without" phenomenon. This phenomenon is characterized by nonspecific early

bacterial lysis caused by cell wall-degrading enzymatic activity of endolysins or phage tail proteins, regardless of whether the phage can successfully replicate in the tested strain. This results in false-positive results and erroneous interpretation of bacterial sensitivity to phages [37]. This can occur when spotting high-titer (i.e., > 10^8 PFU/mL) or undiluted phage lysates. As such, we recommend the use of 1/10 and 1/100 dilutions of a phage lysate for the spot test assay. If lysis occurs only in the undiluted lysate and not for 1/10 or 1/100 dilutions, it might be the result of unspecific "lysis from without." In case of doubt, using the full plaque assay in soft agar overlay is recommended as this technique yields isolated phage plaques, a clear indication of phage replication.

7. A high-throughput strategy can be applied by preparing serial dilutions (10^{-1}, 10^{-2}, and 10^{-3}) of the bacteria to induce in 96-well plates. Using a sterile 96-pin replicator or a multichannel pipet, transfer the diluted bacteria on a rectangular-bottom agar plate and follow the same procedure as for the conventional method. The volume of soft agar overlay will need to be adjusted to 5 mL.

8. Primary phage amplification can also be done using a pure phage stock frozen at −80 °C in 50% glycerol as the seed. In this case, proceed to **step 2** from Subheading 3.6, but instead of adding a phage plug, scrape the surface of the frozen phage glycerol stock with a sterile 1-mL serological pipet to pick up a chunk of frozen phage stock. Melt the frozen chunk directly in a tube containing 5 mL of TY broth + salts = host strain inoculated at 1% from an O/N preculture. Incubate O/N and then proceed to **steps 3–6** from Subheading 3.6. Alternatively, the frozen phage stock can be spotted on top of a soft agar overlay inoculated with the host strain. A lysis zone or phage plaques should be observed on the bacterial lawn after O/N incubation. This method can be useful when the phage infectivity has dropped drastically during storage. The primary amplification can then be performed using a phage plug from the soft agar overlay.

9. With temperate phages, a low multiplicity of infection (MOI) typically promotes amplification through the lytic cycle, while a higher MOI favors lysogeny [38]. An MOI between 0.01 and 0.1 is generally optimal for maximal amplification in broth. Typically, a *C. difficile* culture at an OD_{600} of 0.1 contains approximately 10^7 CFU/mL. In a 10-mL phage amplification broth with ~10^8 CFU, adding 10^6 or 10^7 phages will achieve MOI values of 0.01 and 0.1, respectively. These values are provided as a guideline based on our experience, and the optimal MOI may vary for different phages and hosts.

10. When preparing phage dilutions or mixing suspensions for adsorption assays, it is crucial to mix by gentle inversion and avoiding the use of a vortex. Vortexing can physically damage phage particles and potentially skew adsorption efficiency measurements.
11. A 30-min incubation period is generally sufficient for phage adsorption to occur without allowing progression into later stages of the infection cycle. This duration ensures that most phage particles have time to attach to bacterial cells and inject their DNA, while minimizing the risk of cell lysis or secondary infections that could interfere with adsorption quantification.
12. While spot assay may provide a rapid estimate of phage titers, full-overlay titrations are preferred for adsorption assays because they offer higher accuracy and reproducibility. The complete overlay ensures even distribution of phages and bacteria, reduced edge effects, and facilitates precise plaque counting, which is critical for calculating residual titers after adsorption.
13. To ascertain whether cohesive termini are present in phages, it is possible to compare a non-heated aliquot of a restriction reaction with a heated one. If cohesive termini are present, a band in the non-heated sample will be split into two bands in the heated reaction. If there is no change in profile, using different restriction enzymes might be necessary to determine if cohesive termini are present.
14. It is preferable to use a fresh phage lysate (< 1 week) to maximize particle integrity. Once washed with the ammonium acetate buffer, proceed to observation in the following days.
15. Our experience has shown that TEM grids tend to lose their binding efficacy after several months of storage. It results in low binding of phage particles and poor grid staining. It is therefore recommended to use new grids, or to process them by glow discharge to regenerate their surface and increase wettability.

References

1. Heuler J, Fortier L-C, Sun X (2021) *Clostridioides difficile* phage biology and application. FEMS Microbiol Rev. https://doi.org/10.1093/femsre/fuab012
2. Govind R, Fralick JA, Rolfe RD (2006) Genomic organization and molecular characterization of *Clostridium difficile* bacteriophage ΦCD119. J Bacteriol 188:2568–2577. https://doi.org/10.1128/jb.188.7.2568-2577.2006
3. Selle K, Fletcher JR, Tuson H, Schmitt DS, McMillan L, Vridhambal GS, Rivera AJ, Montgomery SA, Fortier L-C, Barrangou R et al (2020) In vivo targeting of *Clostridioides difficile* using phage-delivered CRISPR-Cas3 antimicrobials. MBio 11:e00019–e00020. https://doi.org/10.1128/mbio.00019-20
4. Camargo AP, Roux S, Schulz F, Babinski M, Xu Y, Hu B, Chain PSG, Nayfach S, Kyrpides NC (2023) Identification of mobile genetic

elements with geNomad. Nat Biotechnol:1–10. https://doi.org/10.1038/s41587-023-01953-y

5. Gebhart D, Williams SR, Bishop-Lilly KA, Govoni GR, Willner KM, Butani A, Sozhamannan S, Martin D, Fortier L-C, Scholl D (2012) Novel high-molecular-weight, R-type bacteriocins of *Clostridium difficile*. J Bacteriol 194:6240–6247. https://doi.org/10.1128/jb.01272-12
6. Sekulovic O, Garneau JR, Néron A, Fortier L-C (2014) Characterization of temperate phages infecting *Clostridium difficile* isolates of human and animal origins. Appl Environ Microb 80:2555–2563. https://doi.org/10.1128/aem.00237-14
7. Fortier L-C, Moineau S (2007) Morphological and genetic diversity of temperate phages in *Clostridium difficile*. Appl Environ Microb 73:7358–7366. https://doi.org/10.1128/aem.00582-07
8. Bearson BL, Brunelle BW (2015) Fluoroquinolone induction of phage-mediated gene transfer in multidrug-resistant salmonella. Int J Antimicrob Agents 46:201–204. https://doi.org/10.1016/j.ijantimicag.2015.04.008
9. Johnson TA, Looft T, Severin AJ, Bayles DO, Nasko DJ, Wommack KE, Howe A, Allen HK (2017) The In-Feed antibiotic carbadox induces phage gene transcription in the swine gut microbiome. mBio 8. https://doi.org/10.1128/mbio.00709-17
10. Fortier L-C, Moineau S (2009) Bacteriophages, methods and protocols, volume 1: isolation, characterization, and interactions. Methods Mol Biol 501:203–219. ISBN 9781588296825
11. Royer ALM, Umansky AA, Allen M-M, Garneau JR, Ospina-Bedoya M, Kirk JA, Govoni G, Fagan RP, Soutourina O, Fortier L-C (2023) *Clostridioides difficile* S-Layer Protein A (SlpA) serves as a general phage receptor. Microbiol Spectrum 11:e03894–e03822. https://doi.org/10.1128/spectrum.03894-22
12. Garneau JR, Depardieu F, Fortier L-C, Bikard D, Monot M (2017) PhageTerm: a tool for fast and accurate determination of phage termini and packaging mechanism using next-generation sequencing data. Sci Rep-uk 7:8292. https://doi.org/10.1038/s41598-017-07910-5
13. Garneau JR, Legrand V, Marbouty M, Press MO, Vik DR, Fortier L-C, Sullivan MB, Bikard D, Monot M (2021) High-throughput identification of viral termini and packaging mechanisms in virome datasets using PhageTermVirome. Sci Rep-uk 11:18319. https://doi.org/10.1038/s41598-021-97867-3
14. Bouras G, Nepal R, Houtak G, Psaltis AJ, Wormald P-J, Vreugde S (2022) Pharokka: a fast scalable bacteriophage annotation tool. Bioinformatics 39. https://doi.org/10.1093/bioinformatics/btac776
15. Seemann T (2014) Prokka: rapid prokaryotic genome annotation. Bioinformatics 30:2068–2069. https://doi.org/10.1093/bioinformatics/btu153
16. McNair K, Zhou C, Dinsdale EA, Souza B, Edwards RA (2019) PHANOTATE: a novel approach to gene identification in phage genomes. Bioinformatics 35:4537–4542. https://doi.org/10.1093/bioinformatics/btz265
17. Terzian P, Ndela EO, Galiez C, Lossouarn J, Bucio REP, Mom R, Toussaint A, Petit M-A, Enault F (2021) PHROG: families of prokaryotic virus proteins clustered using remote homology. NAR Genom Bioinform 3. https://doi.org/10.1093/nargab/lqab067
18. Chan PP, Lin BY, Mak AJ, Lowe TM (2021) tRNAscan-SE 2.0: improved detection and functional classification of transfer RNA genes. Nucleic Acids Res 49:9077–9096. https://doi.org/10.1093/nar/gkab688
19. Laslett D, Canback B (2004) ARAGORN, a program to detect tRNA genes and tmRNA genes in nucleotide sequences. Nucleic Acids Res 32:11–16. https://doi.org/10.1093/nar/gkh152
20. Bland C, Ramsey TL, Sabree F, Lowe M, Brown K, Kyrpides NC, Hugenholtz P (2007) CRISPR recognition tool (CRT): a tool for automatic detection of clustered regularly interspaced palindromic repeats. BMC Bioinform 8:209. https://doi.org/10.1186/1471-2105-8-209
21. Liu B, Zheng D, Jin Q, Chen L, Yang J (2019) VFDB 2019: a comparative pathogenomic platform with an interactive web interface. Nucleic Acids Res 47:D687–D692. https://doi.org/10.1093/nar/gky1080
22. Alcock BP, Raphenya AR, Lau TTY, Tsang KK, Bouchard M, Edalatmand A, Huynh W, Nguyen A-LV, Cheng AA, Liu S et al (2020) CARD 2020: antibiotic resistome surveillance with the comprehensive antibiotic resistance database. Nucleic Acids Res 48:D517–D525. https://doi.org/10.1093/nar/gkz935
23. Bouras G, Grigson SR, Mirdita M, Heinzinger M, Papudeshi B, Mallawaarachchi V, Green R, Kim RS, Mihalia V, Psaltis AJ et al (2025) Protein structure informed bacteriophage genome

annotation with phold. bioRxiv. https://doi.org/10.1101/2025.08.05.668817

24. Heinzinger M, Weissenow K, Sanchez JG, Henkel A, Mirdita M, Steinegger M, Rost B (2024) Bilingual language model for protein sequence and structure. NAR Genom. Bioinform. 6:lqae150. https://doi.org/10.1093/nargab/lqae150
25. Hutson M (2023) Foldseek gives AlphaFold protein database a rapid search tool. Nature. https://doi.org/10.1038/d41586-023-02205-4
26. Boulay A, Leprince A, Enault F, Rousseau E, Galiez C (2025) Empathi: embedding-based phage protein annotation tool by hierarchical assignment. Nat Commun 16:9114. https://doi.org/10.1038/s41467-025-64177-5
27. Jang HB, Bolduc B, Zablocki O, Kuhn JH, Roux S, Adriaenssens EM, Brister JR, Kropinski AM, Krupovic M, Lavigne R et al (2019) Taxonomic assignment of uncultivated prokaryotic virus genomes is enabled by gene-sharing networks. Nat Biotechnol 37:632–639. https://doi.org/10.1038/s41587-019-0100-8
28. Shannon P, Markiel A, Ozier O, Baliga NS, Wang JT, Ramage D, Amin N, Schwikowski B, Ideker T (2003) Cytoscape: a software environment for integrated models of biomolecular interaction networks. Genome Res 13:2498–2504. https://doi.org/10.1101/gr.1239303
29. Gilchrist CLM, Chooi Y-H (2021) Clinker & Clustermap.Js: automatic generation of gene cluster comparison figures. Bioinformatics 37: 2473–2475. https://doi.org/10.1093/bioinformatics/btab007
30. Gaborieau B, Vaysset H, Tesson F, Charachon I, Dib N, Bernier J, Dequidt T, Georjon H, Clermont O, Hersen P et al (2024) Prediction of strain level phage–host interactions across the Escherichia genus using only genomic information. Nat Microbiol 9:2847–2861. https://doi.org/10.1038/s41564-024-01832-5
31. Govind R, Vediyappan G, Rolfe RD, Dupuy B, Fralick JA (2009) Bacteriophage-mediated toxin gene regulation in *Clostridium difficile*. J Virol 83:12037–12045. https://doi.org/10.1128/jvi.01256-09
32. Sekulovic O, Meessen-Pinard M, Fortier L-C (2011) Prophage-stimulated toxin production in *Clostridium difficile* NAP1/027 lysogens. J Bacteriol 193:2726–2734. https://doi.org/10.1128/jb.00787-10
33. Frankowska N, Szarek K, Iwanicki A, Wultańska D, Pituch H, Kabała M, Negri A, Obuchowski M, Hinc K (2025) The Phi027 bacteriophage influences physiology and virulence of the lysogenic strain of *Clostridioides difficile*. Sci Rep 15:18856. https://doi.org/10.1038/s41598-025-04106-0
34. Spinelli S, Veesler D, Bebeacua C, Cambillau C (2014) Structures and host-adhesion mechanisms of lactococcal Siphophages. Front Microbiol 5:3. https://doi.org/10.3389/fmicb.2014.00003
35. Raza T, Andleeb S, Ullah SR, Jamal M, Mehmood K, Ali M (2018) Isolation and characterization of a phage to control vancomycin resistant Enterococcus faecium. Open Life Sci 13:553–560. https://doi.org/10.1515/biol-2018-0066
36. Goh S, Riley TV, Chang BJ (2005) Isolation and characterization of temperate bacteriophages of *Clostridium difficile*. Appl Environ Microbiol 71:1079–1083. https://doi.org/10.1128/aem.71.2.1079-1083.2005
37. Abedon ST (2011) Lysis from Without. Bacteriophage 1:46–49. https://doi.org/10.4161/bact.1.1.13980
38. Zhang M, Zhang T, Yu M, Chen Y-L, Jin M (2022) The life cycle transitions of temperate phages: regulating factors and potential ecological implications. Viruses 14:1904. https://doi.org/10.3390/v14091904

Chapter 6

Preparation of Spores with the 70:30 Method

Eliane de Oliveira Ferreira

Abstract

Clostridioides difficile spores are highly resistant, dormant structures that play a critical role in the transmission and pathogenesis of *C. difficile* infections. The purification of *C. difficile* spores is essential for understanding the mechanisms of spore formation, germination, and the disease. In this chapter, a method for the purification of *C. difficile* spores using a 70:30 medium is described, consisting of 70% SMC (sporulation medium for *Clostridia*) and 30% BHIS (brain–heart infusion supplement). The 70:30 medium is used to induce sporulation and produce high yields of pure *C. difficile* spores. A protocol for the preparation of the 70:30 medium, spore purification, and characterization of the purified spores is described. This method is a valuable tool for researchers studying the biology and pathogenesis of *C. difficile* and will facilitate the development of effective strategies for the prevention and treatment of *C. difficile* infections.

Key words Spore induction, Spore purification, 70:30 method

1 Introduction

Bacterial spores are formed through an asymmetric cell division that results in two distinct cellular compartments. The smaller unit, referred to as the forespore, matures into the spore, while the larger unit, known as the mother cell, contributes to the differentiation process and undergoes autolysis, ultimately releasing the endospore into the extracellular environment [1]. Environmental and intracellular stress signals are sensed by histidine kinase receptors, which trigger the sporulation process by activating Spo0A, a highly conserved transcriptional regulator considered the master regulator of sporulation. Following asymmetric division, gene expression is tightly regulated by four sigma factors. σ^F and σ^E orchestrate early sporulation events in the forespore and mother cell, respectively, until the end of the engulfment stage. They are then replaced by σ^G and σ^K, which drive late sporulation processes [2].

Clostridioides difficile is a major gastrointestinal pathogen in humans and animals, recognized as the primary cause of antibiotic-

Kurt H. Piepenbrink and Jennifer M. Auchtung (eds.), *Clostridioides difficile: Methods and Protocols*, Methods in Molecular Biology, vol. 3046, https://doi.org/10.1007/978-1-0716-5328-9_6,

associated diarrhea. The burden of *C. difficile* infection (CDI) is substantial, resulting in billions of dollars in healthcare costs and over 14,000 annual deaths in the United States alone. Consequently, the organism has been designated an urgent public health threat by the CDC. This anaerobic bacterium survives outside a host by forming a metabolically inactive spore. This complex process creates a dormant structure that shields *C. difficile* from harsh conditions—like oxygen, heat, disinfectants, and desiccation—and grants resistance to antimicrobials [3].

When ingested, the spore germinates (switches back to the vegetative form) upon exposure to bile salts within the host. The resulting vegetative cells then multiply and secrete the toxins that cause disease symptoms. As these cells move through the gastrointestinal tract, unknown signals trigger a return to sporulation. Shed in the feces, these durable spores persist in the environment, presenting a significant challenge for preventing CDI in healthcare settings [4].

To ensure the viability of the future vegetative cell of *C. difficile* derived from the spore, it must be protected from damage to critical components essential for germination and cell growth. Additionally, mechanisms must exist to repair or replace damaged components to prevent cell death [5, 6]. The multiple structural layers of the spore confer resistance to various environmental stresses. From innermost to outermost, these layers include core, inner membrane, germ cell wall, cortex, outer membrane, coat, and exosporium [7]. The core is the innermost compartment of the spore, housing the genomic material and essential enzymes. A key feature contributing to spore resistance is the high concentration of dipicolinic acid (DPA) complexed with calcium ions (Ca^{2+}). This Ca-DPA complex reduces water availability within the spore (to about 25–60% of wet weight), thereby enhancing resistance to a wide range of chemical agents and temperature shifts while also preventing enzyme inactivation [8]. Additionally, Ca-DPA contributes to DNA stabilization against thermal denaturation by intercalating between nitrogenous bases [9]. Another important factor is the abundance of small acid-soluble spore proteins (SASPs), particularly of the α/β type, which provide protection against ultraviolet radiation, moist heat, and oxidative stress caused by hydrogen peroxide (H_2O_2) [10]. These proteins bind to DNA, altering its conformation from B-form to a more compact and resistant A-form, reducing UV-induced pyrimidine dimer formation and heat denaturation [11]. SASPs are degraded during germination, supplying amino acids for early bacterial growth [12]. The core is surrounded by a compressed inner membrane, structurally similar to that of vegetative cells, but with low permeability to small molecules, including water, which further protects the core from harmful agents [13]. The germ cell wall surrounds the inner

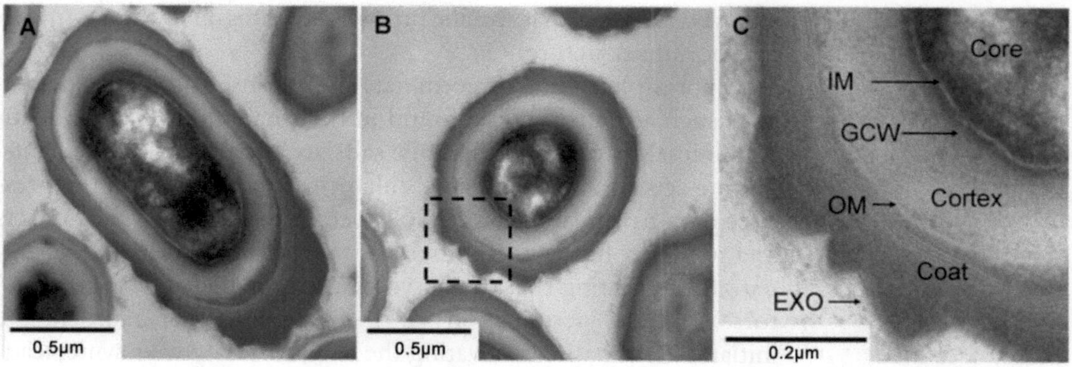

Fig. 1 *Clostridioides difficile* R20291 strain spores structure and cell morphology visualized by transmission electron microscopy (TEM) (**A** to **C**). IM inner membrane, GCW germ cell wall, OM outer membrane, EXO exosporium. (Image adapted from the publication Baloh et al. [14])

membrane and later becomes the cell wall of the vegetative bacterium post-germination [7] (Fig. 1).

Surrounding the germ cell wall is the cortex, composed of thick peptidoglycan with specific modifications not found in vegetative cell walls or the germ cell wall [15]. These modifications are recognized by cortex-lytic enzymes during germination, which selectively degrade the cortex while preserving the germ cell wall [16]. Beyond the cortex lies the outer membrane, derived from the mother cell. Although essential for spore assembly, it does not contribute to resistance [17]. Surrounding this is the spore coat, a protein-rich layer that plays a crucial role in resistance to chemical disinfectants commonly used in hospitals. Studies on *Bacillus* spores suggest similar functions for the spore coat in *Clostridioides* [18]. Additionally, the coat protects the spore from lytic enzymes, chemical agents, and protozoan predation [19].

In *C. difficile* and some *Bacillus* species, an additional outer layer called the exosporium is present. Composed of proteins, including unique glycoproteins, the exosporium contributes to adhesion, hydrophobicity, and spore persistence, especially during recurrent infections [7, 20, 21]. While the exosporium of *C. difficile* was once understudied, recent proteomic analysis has revealed its complex composition, identifying over 180 associated proteins in the outermost layer [22]. Crucially, modern studies have identified key structural and morphogenetic proteins in the *C. difficile* exosporium, including the collagen-like glycoproteins (BclA1, BclA2, and BclA3) and the cysteine-rich morphogenetic proteins (CdeC and CdeM), which are essential for correct assembly and recruitment of other surface proteins [23]. Many of the identified *C. difficile* proteins appear to be adsorbed components or nonstructural cytosolic proteins encased during assembly. However, the identification of morphogenetic factors like CdeC and

CdeM establishes that specific proteins are integral to the *C. difficile* exosporium structure.

Spores of *C. difficile* germinate in response to bile salt derivatives such as taurocholate, found in the small intestine of humans and animals [16, 24]. While bile salts are the primary germinants, glycine acts as a co-germinant, enhancing germination in the presence of taurocholate [25, 26]. Conversely, chenodeoxycholate acts as a potent competitive inhibitor of taurocholate-mediated germination, suggesting that these two bile salts are key in vivo modulators of spore germination [27]. Recognition of taurocholate initiates germination, activating the SleC cortex-lytic enzyme, leading to Ca-DPA release and metabolic reactivation. SleC activation is mediated by serine proteases of the Csp family. Three homologous csp genes have been identified in *C. difficile*, organized in a bicistronic operon: *cspBA* and *cspC*, with *cspB* and *cspA* existing as a gene fusion [28, 29]. Disruption of this operon or catalytic mutations in CspB impair germination, emphasizing its role in SleC activation, though it is not the sole activating protease [26].

Given the *C. difficile* spore's critical role in the organism's life cycle, CDI treatment, and prevention, spore isolation and purification are necessary. This step enables the study of sporulation and germination mechanisms, the investigation of spore properties and resistances, and their use in animal models of CDI.

A primary obstacle about studying *C. difficile* spore morphogenesis was the lack of an assay that reliably induces sporulation in a high percentage of cells. BHIS medium, the standard growth medium for *C. difficile*, failed to promote high-level sporulation in either agar or broth cultures [30]. Although SMC has been reported to efficiently induce sporulation in strain 630 [31], it was observed that different *C. difficile* strains exhibited inconsistent growth rates on this medium. A mixture of 70% SMC medium and 30% BHIS medium, 70:30 medium [32] is the most reliable sporulation medium, with approximately 20% of cells consistently sporulating after 24 h.

A vital consideration when handling *C. difficile* spores is their high tendency to adhere to surfaces (like polypropylene and pipette tips) due to their hydrophobic and anionic nature. To ensure minimal loss and accurate counts, always pipette up and down repeatedly, especially during dilutions. Standard polypropylene is acceptable for short-term, high-concentration stocks (10^7 CFU/mL), but for long-term storage and low-concentration dilutions (10^6 CFU/mL), glass or Teflon-coated tubes must be used to maintain reproducibility. Finally, the inclusion of 1% BSA in the final spore preparation is essential, as it prevents clumping, reduces surface adherence, and facilitates accurate spore enumeration [33].

2 Materials

2.1 *C. difficile* Growth and Sporulation Induction

2.1.1 Growth and Sporulation Media

BHIS broth (37 g brain–heart infusion medium, 5 g yeast extract, 0.1% [wt/vol] L-cysteine) and agar (BHIS broth with 15 g agar per liter) are used for routine cultivation of *C. difficile*. L-Cysteine is typically prepared as a 100× stock (10% wt/vol in water) that can be filter-sterilized and added after autoclaving. When needed, sodium taurocholate can also be prepared as a 100× stock (10% wt/v0l in water) that can be filter-sterilized and added after autoclaving. The 70:30 medium is prepared by combining 63 g/L bacto peptone, 3.5 g/L protease peptone, 1.5 g/L yeast extract, 1.06 g/L Tris base, 0.7 g/L NH_4SO_4, and 11.1 g/L brain–heart infusion. When needed, agar can be added to 15 g/L, with both 70:30 broth and agar sterilized by autoclaving.

2.2 Isolation of *C. difficile* Spores

Lysis buffer: Lysis buffer is prepared filter sterilized by adding 125 mM Tris, 200 mM EDTA, 0.3 mg/mL proteinase K, and 1% sarcosyl to phosphate-buffered saline (1× PBS).

3 Methods

3.1 *C. difficile* Growth and Sporulation Induction

1. Reactivate the *C. difficile* strain from a stock maintained in cryotubes at −80 °C and streak onto BHIS agar plates containing 0.1% sodium taurocholate to induce spore germination.
2. Incubate the plates anaerobically (80% N_2, 10% H_2, and 10% CO_2) for 24 h at 37 °C [31].
3. After 24 h of incubation, transfer a single colony to a tube containing 10 mL of 70:30 medium, as this promotes the sporulation of vegetative cells [34]. Incubate this tube at 37 ° C for 24 h under anaerobic conditions.
4. From this growth, transfer a 1 mL subculture to a new test tube containing 9 mL of 70:30 medium and incubate at 37 °C until the optical density at 600 nm (OD_{600}) reaches approximately 0.5, indicating the logarithmic growth phase of the bacteria [35]. Subsequently, plate 100 μL of this growth onto Petri dishes containing 35 mL of 70:30 medium supplemented with 1.5% agar. Depending on the strain, sometimes it is necessary incubate the plates at least 10 days.
5. To visualize the spore, there are some techniques that can be used, but the Schaeffer–Fulton stain [36], also known as Wirtz–Conklin stain, with malachite green dye is simple and easy to observe the spores under an optical microscope (magnification: 1000×). After preparing the slides, spores will appear in green and the vegetative cells in red.

6. For counting the number of spores produced by each strain (spores/mL), after growing the cells in the 70:30 media (**item 2.2.1**), you can prepare serial dilutions in 1X phosphate buffered saline (PBS), plate dilutions onto BHIS agar plates with 0.1% taurocholate, incubate for at least 72 h, and count resulting colonies (*see* **Note 1**). An alternative approach for counting spores, which was previously described, is to prepare samples for confocal observations and counting [37].

3.2 Isolation of C. difficile Spores

1. After obtaining the growth in 70:30 plates (**item 2.2.1**) to obtain spores, scrape the plates with by using microbiological loops and transfer the content to tubes containing 5 mL of sterile ice water and kept them at 4 °C for approximately 16 h [31].
2. After this period, wash the total scraped content of the plates twice with 1 mL of sterile ice water and then resuspend them in 1 mL of lysis buffer. Incubate under gentle agitation for 2 h at 37 °C (*see* **Note 2**).
3. After incubation, centrifuge the tubes (8000 × g for 10 min), and resuspend the sediment with 1 mL of sterile ice water and wash 5 to 10 times.
4. After the last wash, incubate the suspended contents at 60 °C for 20 min for a second lysis of the residual vegetative cells.
5. Store aliquots containing 1 mL of each strain at 4 °C until use [31, 38].
6. To separate and recover spores, apply the entire 1-mL spore preparation slowly to the top of 4 mL of 50% sucrose gradient in polypropylene conical tubes [27].
7. Centrifuge gradients at 4500 × g for 15 min at 25 °C in a swinging bucket rotor.
8. After centrifuging, the vegetative cells and the debris will be at the interface phase and the spores will form a pellet at the bottom.
9. Add 1 mL of sterile ice-cold water to the sediment and transfer the suspension to new tubes.
10. Centrifuge at 16,000 × g for 2 min at room temperature and decant the supernatant.
11. Repeat washing steps several times (five washes in sterile H_2O and three washes in 1× PBS + 1% BSA).
12. Storing spores: for optimal stability, keep the final working spore stock at room temperature using glass or Teflon-coated tubes (*see* **Note 3**).

4 Notes

1. Some strains take a little bit more time to germinate and grow to colonies than others, depending on their ability to produce spores. So, sometimes it is necessary to leave plates incubate a little longer before counting.
2. Lysis buffer lyses the vegetative cells and release the endospores from their interior.
3. Be aware that spore viability is not constant; it varies by strain and stock. Therefore, any stock older than 2 weeks should be retested for viability. Before using the stock for any plating or experiment, you must heat the spore preparation to 55 °C for 20 min to prepare the spores for subsequent steps [39].

References

1. Jensen GJ (2017) HHS Public Access 14:535–542. https://doi.org/10.1038/nrmicro.2016.85.Sporulation
2. Pereira FC, Saujet L, Tome AR, Serrano M, Monot M, Couture-Tosi E, Martin-Verstraete-I, Dupuy B, Henriques AO (2013) The spore differentiation pathway in the enteric pathogen Clostridium difficile. PLoS Genet 9:e1003782. https://doi.org/10.1371/journal.pgen.1003782
3. Bloomfield LE, Riley TV (2016) Epidemiology and risk factors for community-associated Clostridium difficile infection: a narrative review. Infect Dis Ther 5:231–251. https://doi.org/10.1007/s40121-016-0117-y
4. Kumar N, Browne HP, Viciani E, Forster SC, Clare S, Harcourt K, Stares MD, Dougan G, Fairley DJ, Roberts P, Pirmohamed M, Clokie MRJ, Jensen MBF, Hargreaves KR, Ip M, Wieler LH, Seyboldt C, Norén T, Riley TV, Kuijper EJ, Wren BW, Lawley TD (2019) Adaptation of host transmission cycle during Clostridium difficile speciation. Nat Genet 51:1315–1320. https://doi.org/10.1038/s41588-019-0478-8
5. Nicholson WL, Munakata N, Horneck G, Melosh HJ, Setlow P (2000) Resistance of bacillus endospores to extreme terrestrial and extraterrestrial. Environments 64:548–572
6. Paredes-Sabja D, Bond C, Carman RJ, Setlow P, Sarker MR (2008) Germination of spores of Clostridium difficile strains, including isolates from a hospital outbreak of Clostridium difficile-associated disease (CDAD). Microbiology 154:2241–2250. https://doi.org/10.1099/mic.0.2008/016592-0
7. Pizarro-Guajardo M, Calderón-Romero P, Castro-Córdova P, Mora-Uribe P, Paredes-Sabja D, Calderon-Romero P, Castro-Cordova P, Mora-Uribe P, Paredes-Sabja D (2016) Ultrastructural variability of the exosporium layer of Clostridium difficile spores. Appl Environ Microbiol 82:2202–2209. https://doi.org/10.1128/AEM.03410-15
8. Pizarro-Guajardo M, Olguin-Araneda V, Barra-Carrasco J, Brito-Silva C, Sarker MR, Paredes-Sabja D (2014) Characterization of the collagen-like exosporium protein, BclA1, of Clostridium difficile spores. Anaerobe 25: 18–30. https://doi.org/10.1016/j.anaerobe.2013.11.003
9. Setlow P (2007) I will survive: DNA protection in bacterial spores. Trends Microbiol 15:14–16. https://doi.org/10.1016/j.tim.2007.02.004
10. Wetzel D (2015) Small acid-soluble spore proteins of Clostridium acetobutylicum are able to protect DNA in vitro and are specifically cleaved by germination protease GPR and spore protease YyaC:2098–2109. https://doi.org/10.1099/mic.0.000162
11. Douki T (2006) Effect of denaturation on the photochemistry of pyrimidine bases in isolated. DNA 82:45–52. https://doi.org/10.1016/j.jphotobiol.2005.08.009
12. Traag BA, Pugliese A, Setlow B, Setlow P, Losick R (2013) A conserved ClpP-like protease involved in spore outgrowth in Bacillus subtilis. 90:160–166. https://doi.org/10.1111/mmi.12355
13. Cowan AE, Olivastro EM, Koppel DE, Loshon CA, Setlow B, Setlow P (2004) Lipids in the inner membrane of dormant spores of Bacillus

species are largely immobile. Proc Natl Acad Sci USA 101:7733–7738. https://doi.org/10.1073/pnas.0306859101
14. Baloh M, Nerber HN, Sorg JA (2022) Imaging Clostridioides difficile spore germination and germination proteins. J Bacteriol 204(7):1–14. e00210-22
15. Tehri N, Kumar N, Raghu HV, Vashishth A (2018) Biomarkers of bacterial spore germination. Ann Microbiol:513–523
16. Paredes-Sabja D, Sarker MR (2011) Germination response of spores of the pathogenic bacterium Clostridium perfringens and Clostridium difficile to cultured human epithelial cells. Anaerobe 17:78–84. https://doi.org/10.1016/j.anaerobe.2011.02.001
17. Paredes-Sabja D, Shen A, Sorg JA (2014) Clostridium difficile spore biology: sporulation, germination, and spore structural proteins. Trends Microbiol 22:406. https://doi.org/10.1016/J.TIM.2014.04.003
18. Piktel E, Pogoda K, Roman M, Niemirowicz K, Tokajuk G, Wróblewska M, Szynaka B, Kwiatek WM, Savage PB, Bucki R (2017) Sporicidal activity of ceragenin CSA-13 against Bacillus subtilis. Nat Publ Gr:1–12. https://doi.org/10.1038/srep44452
19. Klobutcher LA, Ragkousi K, Setlow P (2015) The Bacillus subtilis spore coat provides '"eat resistance"' during phagocytic predation by the protozoan Tetrahymena thermophila. Proc Natl Acad Sci USA. https://doi.org/10.1073/pnas.0507121102
20. Leggett MJ, Setlow P, Sattar SA, Maillard J-Y (2016) Assessing the activity of microbicides against bacterial spores: knowledge and pitfalls. J Appl Microbiol 120:1174–1180. https://doi.org/10.1111/jam.13061
21. Strong PCR, Fulton KM, Aubry A, Foote S, Twine SM, Logan SM (2014) Identification and characterization of glycoproteins on the spore surface of Clostridium difficile. J Bacteriol 196:2627–2637. https://doi.org/10.1128/JB.01469-14
22. Diaz-Gonzalez F, Milano M, Olguin-Araneda-V, Pizarro-Cerda J, Castro-Cordova P, Tzeng S-C, Maier CS, Sarker MR, Paredes-Sabja D (2015) Protein composition of the outermost exosporium-like layer of Clostridium difficile 630 spores. J Proteome 123:1–13. https://doi.org/10.1016/j.jprot.2015.03.035
23. Barra-Carrasco J, Olguin-Araneda V, Plaza-Garrido A, Miranda-Cardenas C, Cofre-Araneda G, Pizarro-Guajardo M, Sarker MR, Paredes-Sabja D (2013) The Clostridium difficile exosporium cysteine (CdeC)-rich protein is required for exosporium morphogenesis and coat assembly. J Bacteriol 195:3863–3875. https://doi.org/10.1128/JB.00369-13
24. Sorg JA, Sonenshein AL (2008) Bile salts and glycine as cogerminants for Clostridium difficile spores. J Bacteriol 190:2505–2512. https://doi.org/10.1128/JB.01765-07
25. Francis MB, Allen CA, Shrestha R, Sorg JA (2013) Bile acid recognition by the Clostridium difficile germinant receptor, CspC, is important for establishing infection. PLoS Pathog 9:e1003356. https://doi.org/10.1371/journal.ppat.1003356
26. Kevorkian Y, Shirley DJ, Shen A (2016) Regulation of Clostridium difficile spore germination by the CspA pseudoprotease domain. Biochimie 122:243–254. https://doi.org/10.1016/j.biochi.2015.07.023
27. Sorg JA, Sonenshein AL (2010) Inhibiting the initiation of Clostridium difficile spore germination using analogs of chenodeoxycholic acid, a bile acid. J Bacteriol 192:4983–4990. https://doi.org/10.1128/JB.00610-10
28. Burns DA, Heap JT, Minton NP (2010) SleC is essential for germination of Clostridium difficile spores in nutrient-rich medium supplemented with the bile salt taurocholate. J Bacteriol 192:657–664. https://doi.org/10.1128/JB.01209-09
29. Baloh M, Sorg JA (2023) Clostridioides difficile spore germination: initiation to DPA release. Curr Opin Microbiol 65:101–107. https://doi.org/10.1016/j.mib.2021.11.001
30. Sorg JA, Dineen SS (2009) Laboratory maintenance of Clostridium difficile. Curr Protoc Microbiol Chapter 9:Unit9A.1. https://doi.org/10.1002/9780471729259.mc09a01s12
31. Permpoonpattana P, Tolls EH, Nadem R, Tan S, Brisson A, Cutting SM (2011) Surface layers of Clostridium difficile endospores. J Bacteriol 193:6461–6470. https://doi.org/10.1128/JB.05182-11
32. Putnam EE, Nock AM, Lawley TD, Shen A (2013) SpoIVA and SipL are Clostridium difficile spore morphogenetic proteins. J Bacteriol 195:1214–1225. https://doi.org/10.1128/JB.02181-12
33. Edwards AN, Nawrocki KL, McBride SM (2014) Conserved oligopeptide permeases modulate sporulation initiation in Clostridium difficile. Infect Immun 82:4276–4291. https://doi.org/10.1128/IAI.02323-14
34. Adams CM, Eckenroth BE, Putnam EE, Doublie S, Shen A (2013) Structural and functional analysis of the CspB protease required for clostridium spore germination. PLoS Pathog 9:e1003165. https://doi.org/10.1371/journal.ppat.1003165

35. Permpoonpattana P, Phetcharaburanin J, Mikelsone A, Dembek M, Tan S, Brisson M-C, La Ragione R, Brisson AR, Fairweather N, Hong HA, Cutting SM (2013) Functional characterization of Clostridium difficile spore coat proteins. J Bacteriol 195:1492–1503. https://doi.org/10.1128/JB.02104-12
36. Szulc M, Schein S, Silva FVM, Opretzka J, Benedikt J (2017) The Bacterial endospore stain on Schaeffer Fulton using variation of methylene blue solution. J Phys. https://doi.org/10.1088/1742-6596/755/1/011001
37. Edwards AN, McBride SM (2014) Initiation of sporulation in Clostridium difficile: a twist on the classic model. FEMS Microbiol Lett 358: 110–118. https://doi.org/10.1111/1574-6968.12499
38. Lawley TD, Croucher NJ, Yu L, Clare S, Sebaihia M, Goulding D, Pickard DJ, Parkhill J, Choudhary J, Dougan G, Acteriol JB (2009) Proteomic and genomic characterization of highly infectious. J Bacteriol 191: 5377–5386. https://doi.org/10.1128/JB.00597-09
39. Edwards AN, Mcbride SM (2016) HHS Public Access:117–128. https://doi.org/10.1007/978-1-4939-6361-4

Chapter 7

An Integrated Workflow for the Purification and Fluorescence Imaging of *Clostridioides difficile* Spores Using Lanthanide Beta-Diketone Complexes

Ajitha Sundaresan and Ian Cheong

Abstract

We present a robust protocol for both the purification and imaging of *C. difficile* spores. First, we describe a simplified, one-step isopycnic centrifugation method using Percoll that enables efficient separation of spores from vegetative cells. Next, we introduce a fluorescence-based imaging method using lanthanide beta-diketone complexes (primarily $Eu[TTA]_3$) that allows for the differential visualization of dormant and germinated spores, as well as sporulating cells. These complexes exploit the unique photophysical properties of lanthanides, including sharp emission spectra and resistance to photobleaching, to enhance spore imaging under USB-powered, wide-field, and confocal microscopy. Staining is instantaneous, and no permeabilization or fixation steps are required. We further demonstrate co-staining with dyes such as DAPI and FITC–dextran to reveal spore compartmentalization and confirm germination status. This integrated approach enables detailed characterization of *C. difficile* spore morphology and offers a versatile way to study spore germination, antibiotic response and pathogen persistence at single-spore resolution.

Key words Lanthanide beta-diketones, Live imaging, *C. difficile* spores, Fluorescence microscopy, Confocal microscopy

1 Introduction

Clostridioides difficile is a spore-forming, gram-positive anaerobic pathogen and a leading cause of hospital-associated infections worldwide [1]. The clinical burden of *C. difficile* infection (CDI) is exacerbated by high recurrence rates, resistance to antibiotics, and the ability to persist in healthcare environments. These characteristics are largely attributable to its capacity to form spores [2]. Spore formation is central to the *C. difficile* life cycle. They are metabolically dormant and resistant to environmental stressors, but germinate rapidly under favorable conditions to reestablish infection. Germination is therefore a pivotal control point in path-

Kurt H. Piepenbrink and Jennifer M. Auchtung (eds.), *Clostridioides difficile: Methods and Protocols*, Methods in Molecular Biology, vol. 3046, https://doi.org/10.1007/978-1-0716-5328-9_7,

ogenesis [3], and understanding the molecular and structural changes associated with this process is essential for developing targeted therapeutic strategies.

Traditional methods for studying spore germination often rely on bulk measurements (e.g., optical density changes [4, 5] and Ca-DPA release [6, 7]) which offer only population-level insights and lack the resolution to distinguish individual spore states. Phase-contrast microscopy [8], while useful for identifying phase-bright (dormant) versus phase-dark (germinated) spores, suffers from limited specificity and contrast, especially when spores are aggregated and overlapping within the visual field.

To address these limitations, we introduce a fluorescence-based imaging approach that exploits the unique photophysical properties of lanthanide beta-diketone complexes. Lanthanides such as europium (Eu^{3+}), terbium (Tb^{3+}), and samarium (Sm^{3+}) form stable coordination complexes with beta-diketones, producing sharp, long-lived emission signals that are resistant to photobleaching [9]. When applied to *C. difficile* spores, these complexes allow for the enhanced visualization of spores during germination and sporulation. In particular, the $Eu[TTA]_3$ complex provides strong contrast between dormant and germinated spores [10] under USB-powered, wide-field, and confocal microscopy.

In parallel, we present a simple and scalable method for the purification of bacterial spores using isopycnic centrifugation. We have used this purification method for purifying a wide variety of *Bacillus* and *Clostridium* spores in all our previous work [5, 7, 10–13]. Unlike gradient layering protocols [14, 15], this one-step procedure forms a density gradient in situ, simplifying spore purification while maintaining high purity. Together, these methodologies enable the rapid preparation and high-resolution imaging of *C. difficile* spores, creating new opportunities to investigate germination dynamics and spore viability and to evaluate the efficacy of candidate germination inhibitors.

2 Materials

2.1 Sporulation Media

- *BHIS medium (BHIS media)*: Dissolve 37 gm Brain–Heart infusion broth (237500, BD Difco), 5 gm yeast extract (LP0021, Oxoid), and 1 gm L-cysteine (C7352, Sigma) in 1 L water. Autoclave at 121 °C, 20 min, 50% exhaust and cool the solution to room temperature. Store the media at 4 °C until further use.

- *0.1% taurocholate–BHIS medium (TCA-BHIS media)*: Prepare 1 L of BHIS media as described above. Autoclave at 121 °C, 20 min, 50% exhaust. Filter sterilize 10 mL of 10% (w/v) sodium taurocholate (TCA) (86339, Sigma) using a 0.2-μm filter (Sartorius Minisart®). Add the filtered TCA solution to the autoclaved media after cooling the solution to room temperature. Store the media at 4 °C until further use.

- *BHIS medium agar (BHIS agar)*: Prepare 1 L of BHIS media as described above and add 15 gm Bacto agar (214010, BD Difco) to the solution. Autoclave at 121 °C, 20 min, 50% exhaust and pour into petri dishes (20–25 mL per dish) after cooling. Store the plates at 4 °C until further use.
- *0.1% taurocholate–BHIS medium agar (TCA-BHIS agar)*: Prepare 1 L of BHIS agar solution as described above and autoclave at 121 °C, 20 min, 50% exhaust. Filter sterilize 10 mL of 10% (w/v) TCA solution using a 0.2-μm filter (Sartorius Minisart®). Add the filtered TCA solution to the autoclaved solution after cooling. Pour into petri dishes (20–25 mL per dish) and store the plates at 4 °C until further use.

2.2 Spore Harvest Solutions

1X Phosphate buffer saline (PBS).

Percoll: (17089109, GE healthcare) stored at 4 °C.

2.3 Germinant Solution

Germinant mix: Prepare stock solutions of TCA (20 mM) and glycine (G7126, Sigma, 100 mM). Combine 250 μL of TCA solution, 250 μL of glycine solution, and 10 μL of Oxyrase for broth (SAE0013, Sigma). Final concentrations of TCA and glycine in the germinant solution are 10 mM and 49 mM, respectively.

2.4 Fluorescent Dyes

- *Lanthanides*: Europium (III) chloride hexahydrate (212881, Sigma), Terbium (III) chloride hexahydrate (212903, Sigma), Samarium (III) chloride hexahydrate (204277, Sigma). *Storage*: Since lanthanide salts absorb moisture easily, storing them in a dry box will help keep them dry and stable.
- *Beta-diketones*: TTA/2-thenoyltrifluoroacetone (T27006, Sigma), NTFA/1-(2-napthoyl)-3,3,3-trifluoroacetone (343633, Sigma), BTFA/4,4,4-trifluoro-1-phenyl-1,3-butanedione (217042, Sigma).
- *Lanthanide–TTA complex stock solutions*: Mix equal volumes (200 μL) of lanthanide chloride (3 mM) and TTA (9 mM) solutions in 3% (v/v) ethanol. Similarly, prepare other europium beta-diketone complexes using europium and beta-diketone solutions in 3% (v/v) ethanol. The only exception to this is in the case of the NTFA complex, where a higher amount of ethanol (50%) is required for solubilizing NTFA. Vortex the mixed solutions at maximum speed for 30 min to aid complexation. Final stock concentrations of lanthanide beta-diketones are 1.5 mM. For co-staining assays, use double the concentration of $Eu[TTA]_3$ (3 mM), so that the final concentration remains 1.5 mM after dilution with the other dye. Lanthanide beta-

diketone dye solutions work best when freshly prepared though they still provide sufficient signal when stored at room temperature for a few days.

- *FITC-dextran:* FITC-dextran 3–5 kDa (FD4, Sigma)—Prepare 1 mg/mL stock solution in water and use it as such. Store this stock solution at −20 °C for repeated use.
- *DAPI:* DAPI (D9542, Sigma): Prepare 1 mM solution of DAPI in DMSO and serially dilute it in 3% ethanol to obtain a final stock concentration of 5 μM. Store this stock solution at −20 °C for repeated use.

2.5 Microscopy Materials

- *Microscope slides:* Superfrost microscope slides (cat no. FIS12-550-15, Fisher Scientific).
- *Coverslips:* #1.5 square coverslips (cat no. VWRI631-0125 22 mm × 22 mm, VWR).
- *Nail polish.*
- *Immersion oil.*

3 Methods

3.1 C. difficile Preparation

3.1.1 Sporulation

The sporulation protocol below is adapted from multiple sources [10, 16, 17]. We usually sporulate *C. difficile* (ATCC 700057) on solid media, but the entire protocol (including isopycnic gradient purification) works equally well with sporulation in liquid media.

1. Prior to the day of inoculation, pre-equilibrate 10 BHIS agar plates (for sporulation), 1 TCA-BHIS agar plate (for inoculation), and TCA-BHIS media (5 mL in a conical tube) inside the anaerobic chamber. We use a Plas Labs anaerobic chamber connected with a triple gas cylinder (10% CO_2, 10% H_2, 80% N_2). Face the plates upwards to prevent water condensation on the agar surface. *Important*: Ensure that the anaerobic chamber is maintained at <0.5% oxygen for the entire duration of *C. difficile*'s growth and sporulation.
2. On the day of inoculation, streak approximately 20 μL of *C. difficile* glycerol stock on a TCA-BHIS agar plate and incubate overnight. *Recommended practice*: Swab the anaerobic chamber gloves with isopropanol prior to handling any biological sample. Switch off all airflow generating devices (e.g., fans) inside the chamber while handling the biological material to prevent risk of contamination.
3. Next morning, inoculate one colony from the agar plate into the 5 mL TCA-BHIS media using a 10-μL inoculating loop.

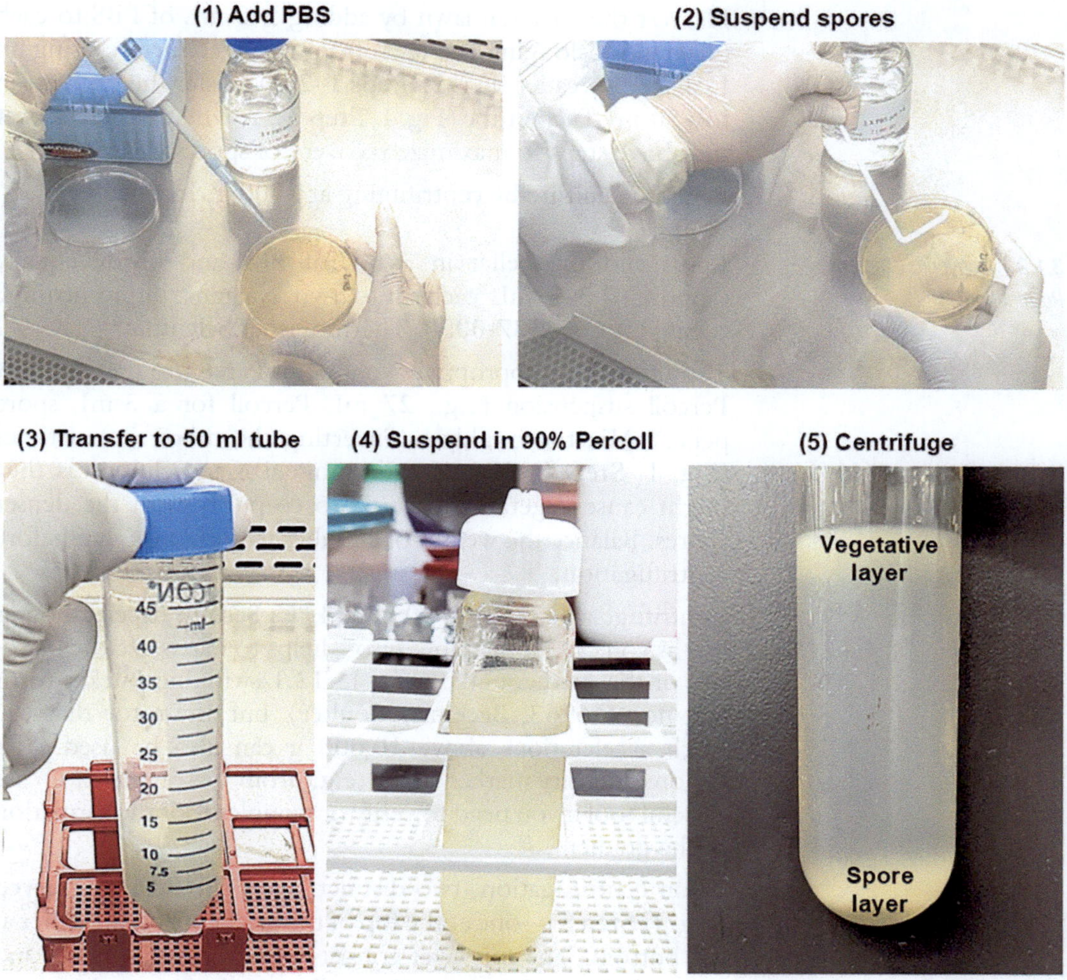

Fig. 1 *C. difficile* spore purification using Percoll isopycnic centrifugation. (1–5) show the sequential steps involved in separating phase bright spores from vegetative bacteria in a sporulating culture grown on BHIS agar plates. Percoll gradients are formed in situ, and bacterial forms migrate to their equilibrium positions based on buoyant density. Vegetative bacteria, being lighter, accumulate at the top, while heavier spores settle at the bottom of the gradient

4. After approximately 6 h when the culture has reached an O.D. of 0.3–0.4, spread 250 μL of this culture onto BHIS agar plates using an L-shaped spreader to initiate sporulation. Our usual practice is to perform a gas purge of the anaerobic chamber every alternate day.
5. Six days later, transfer the agar plates from the anaerobic chamber to a Class II biosafety cabinet (BSC). The plates should now be covered by a fully grown bacterial lawn consisting of spores and vegetative bacteria.

6. Harvest the bacterial lawn by adding 1–2 mL of PBS to each plate and dislodging the lawn from the agar surface using an L-shaped spreader (Fig. 1, **Steps 1–2**). Transfer the suspension to a 50-mL falcon tube (Fig. 1, **Step 3**). You may wish to repeat the above steps to maximize recovery of spores.
7. Pellet the culture by centrifuging at 3900 g, 15 min, 4 °C.

3.1.2 Isopycnic Gradient Purification

1. Resuspend the pellet in 5 mL of PBS and divide equally (approx. 2.5–3 mL each) into two Nalgene ultracentrifuge tubes (cat no. 3137-0050, ThermoFisher Scientific).
2. Top up with an appropriate volume of Percoll to create a 90% Percoll suspension (e.g., 27 mL Percoll for a 3-mL spore pellet). Mix thoroughly by inverting the tubes several times (Fig. 1, **Step 4**). Vortex to homogenize any clumps as they might cause vegetative bacteria to co-purify with the denser spores. Balance the weight of all tubes to within 0.5 gm before centrifugation.
3. Centrifuge at 15,000 g, 30 min, 4 °C. Setting deceleration as low as you can—preferably zero—helps to maximize the integrity of the gradient. We use a JS-13.1 swinging bucket rotor (cat no. 346963, Beckman Coulter), but any rotor that will reach accelerations above 10,000 g can also be used. *Tip*: Colored density marker beads (e.g., from Cospheric LLC) are a useful tool if you need to verify or troubleshoot the formation of the gradient.
4. Upon centrifugation, two distinct layers should be observed (Fig. 1, **Step 5**)—one at the top of the gradient and the other at the bottom. The top layer consists of vegetative bacteria and can resemble a sticky mass, depending on the type and amount of bacteria. The spores, on the other hand, are far denser than the vegetative cells and form the bottom layer. This bottom layer is not a tight pellet, so care should be taken when handling the tube. The zone in between the top and bottom layers should be relatively transparent, and if it appears turbid, it could mean that the isopycnic gradient did not form properly, possibly due to the gradient being overloaded with excessive biological material. If this happens, split the Percoll suspension into two tubes, top each up with Percoll, and repeat the isopycnic centrifugation. The mid-zone should now be clear.
5. *Critical step*: Carefully aspirate the Percoll above the bottom spore layer using a Pasteur pipette attached to a vacuum. Alternatively, a 1-mL serological pipette with its cotton filter removed also works well. To avoid aspirating the spores by accident, do not aspirate too close to the bottom spore layer.

6. *Critical step*: Carefully transfer the bottom spore layer to a 50-mL conical tube using a serological pipette. Avoid touching the walls of the Nalgene tube since vegetative forms often adhere to the sides of the tube.
7. Resuspend the pellet with PBS and centrifuge at 3900 g, 15 min, 4 °C.
8. Aspirate the supernatant and resuspend the spore pellet in 2 mL of PBS.
9. Confirm the spore fraction quality by imaging under phase contrast microscopy to ensure >99% phase bright spores.
10. Count the spore concentration using a Neubauer counting chamber (Petroff Marenfield) with special depth and adjust the spore count to approximately to the order of 10^9 CFU/ml by diluting with PBS.

3.1.3 C. difficile *Spore Germination*

1. To 35 μL of dormant *C. difficile* spores, add 500 μL of germinant solution in a 1.5-mL microfuge tube. Final concentrations of TCA and glycine after spore addition are approximately 9 mM and 45 mM, respectively. These concentrations typically produce a 70–80% germination rate. *Note*: Oxyrase is used to maintain anaerobicity when germination experiments need to be performed outside the anaerobic chamber.
2. Incubate the germinating spore suspension at 37 °C without shaking.
3. After 30 min, centrifuge the tube at 3900 g, 5 min.
4. Discard the supernatant and resuspend in PBS (35 μL) containing Oxyrase (2 μL) for downstream steps. *Note*: Oxyrase can be omitted if viable outgrowth from spores is not required.

3.1.4 C. difficile *Sporulating Culture*

1. Prior to the day of inoculation, pre-equilibrate a TCA-BHIS agar plate, BHIS media (20 mL), and TCA-BHIS media (5 mL) in the anaerobic chamber.
2. On the day of inoculation, streak approximately 20 μL of *C. difficile* glycerol stock on a TCA-BHIS agar plate and incubate overnight.
3. Next morning, inoculate one colony from the agar plate into the 5 mL TCA-BHIS media using a 10-μL inoculating loop.
4. After approximately 6 h when the culture has reached an O.D. of 0.3–0.4, transfer 500 μL into 20 mL of BHIS media to initiate sporulation.
5. After 2 days, remove the sporulating culture from the anaerobic chamber and transfer to a BSC.

6. Centrifuge culture at 3900 g, 15 min.
7. Remove the media supernatant and resuspend in 500 μL of PBS to use for further downstream steps. *Note*: Add Oxyrase to maintain viability of sporulating culture.

3.2 Sample Preparation for Imaging

3.2.1 Staining with Eu [TTA]$_3$ and Other Lanthanide Beta-Diketones

1. Aliquot 3.5 μL of *C. difficile* sample (dormant/germinated spores or sporulating rods) into a 1.5-mL microfuge tube.
2. *Considerations for interference*: The lanthanide beta-diketone dye signal can be affected by interfering factors from the sample matrix, such as competing lanthanides, complexing anions and Oxyrase. Whenever such interactions exist, it is recommended to resuspend in a suitable buffer (e.g., PBS, HEPES or MES) which minimizes interference. In cases where the interacting factors cannot be removed, optimization of sample/dye ratio might be required.
3. To the sample, add 50 μL of freshly prepared Eu[TTA]$_3$ (1.5 mM) or other lanthanide beta-diketone combination so that the final concentration is 1.4 mM.
4. *Solvent considerations*: The solvent used for lanthanide and beta-diketone dissolution can also affect the dye signal. A higher organic solvent concentration improves dye signal but can compromise the spore structure and viability especially in live cell germination tracking. A 3% ethanol solvent concentration is suitable for most lanthanide beta-diketones except NTFA, where a higher ethanol concentration is required.
5. Mix properly using a pipette.
6. Mount 10 μL of stained spore sample on a microscope glass slide with a coverslip.
7. Seal the coverslip with nail polish. *Note*: When using higher concentrations of ethanol in the solvent (e.g. 50%), it is recommended to image the slides as soon as possible to avoid sample drying effects.

3.2.2 Co-Staining with Other Dyes

1. Aliquot 3.5 μL of *C. difficile* sample (dormant/germinated spores or sporulating rods) into a 1.5-mL microfuge tube.
2. To this, add 25 μL of DAPI or FITC–dextran (3–5 kDa) and 25 μL of freshly prepared Eu[TTA]$_3$ (1.5 mM) dye. Final dye concentrations after spore addition are 2.5 μM (DAPI), 0.45 mg/mL (FITC dextran 3–5 kDa), and 1.4 mM (Eu [TTA]$_3$). *Tip*: For staining with DAPI, an incubation time of 30 min prior to addition of Eu[TTA]$_3$ helps improve staining.
3. Mix properly using a pipette.

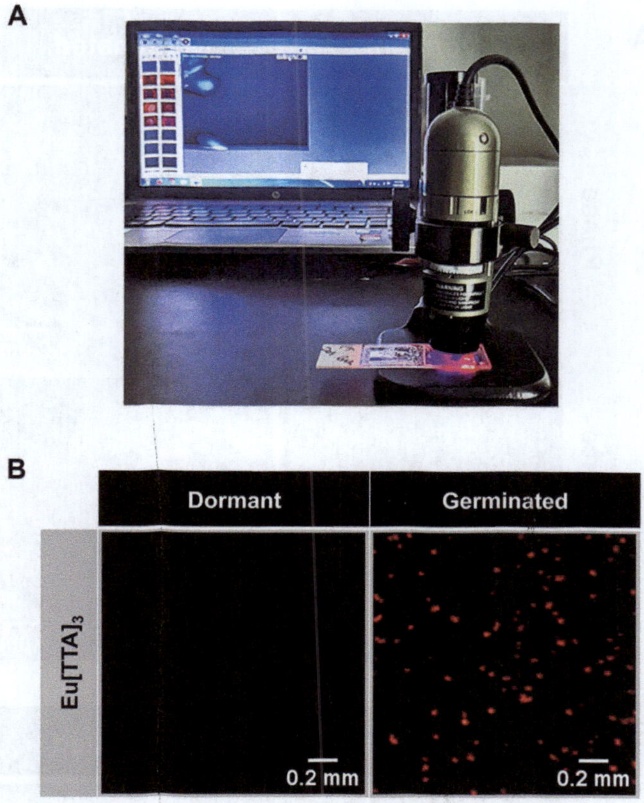

Fig. 2 *C. difficile* spore identification using Eu[TTA]$_3$ staining and USB-powered microscopes. (**a**) Image of the Dinolite™ Premier Digital microscope with UV excitation, connected to a laptop via USB. A sample containing Eu[TTA]$_3$-stained *C. difficile* spores is shown fluorescing red upon UV excitation. (**b**) Fluorescence images of Eu[TTA]$_3$-stained dormant and germinated *C. difficile* spores acquired using the Dinolite™ microscope. Only germinated spores fluoresce under these imaging conditions. Scale bar 0.2 mm

4. Mount 10 μL of stained spore samples on a microscope glass slide with a coverslip.
5. Seal coverslip with nail polish.

3.3 Microscopy

3.3.1 USB-Powered Microscopy

1. USB microscopes with fluorescence imaging capabilities are sufficient for imaging *C. difficile* spores stained with Eu [TTA]$_3$. The USB-powered AM4113T DinoLite™ Premier digital microscope is presented here as an example (Fig. 2a).
2. Capture images at 55× magnification in UV fluorescence mode using the Dino Capture v2.0 software.
3. Dormant *C. difficile* spores do not stain very well under the UV excitation of this microscope and are hence difficult to visualize. In comparison, germinated spores exhibit red fluorescence (Fig. 2b).

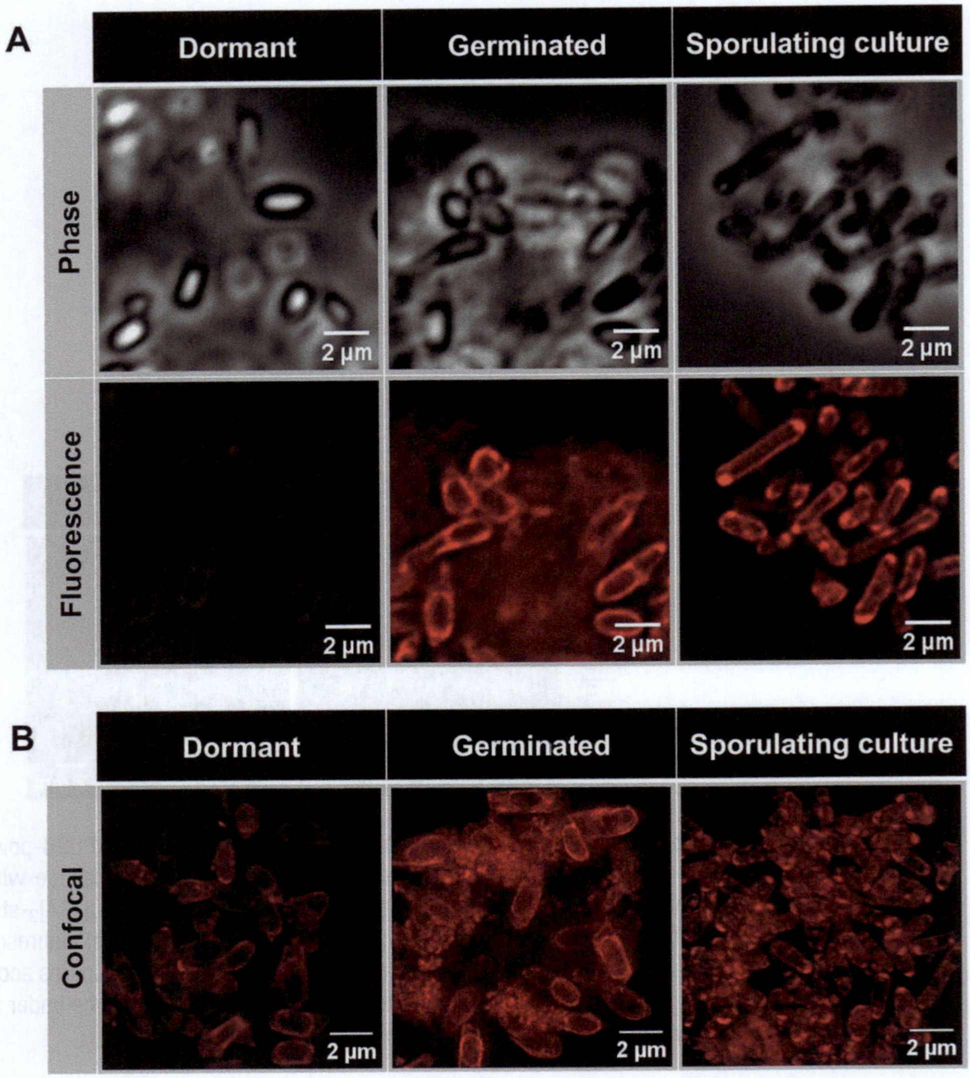

Fig. 3 *C. difficile* spore imaging using Eu[TTA]$_3$ staining and higher-resolution microscopes. (**a**) Wide-field images of Eu[TTA]$_3$-stained dormant and germinated spores, as well as a sporulating culture. Phase contrast (top) and fluorescence (bottom) images are shown. Dormant spores, though less fluorescent than germinated spores and sporulating rods, remain visible. (**b**) Confocal fluorescence images of Eu[TTA]$_3$-stained dormant and germinated spores, as well as a sporulating culture. Dormant spores are distinctly visible, with outer layers stained similarly to germinated spores. All fluorescence images are pseudocolored. Scale bar 2 μm

4. While the low resolution of this microscope does not permit the visualization of spore morphology, it can still nonetheless be used for rapid triage of suspected spore material in the field before further confirmatory tests in a central laboratory.

3.3.2 Wide-Field Microscopy

1. *C. difficile* morphology can be easily visualized using a fluorescence wide-field microscope like the Zeiss Axio Observer 7 inverted microscope. Using higher magnification lenses of 63× or 100× (we typically use a Plan Apochromat 100×/1.4 Oil Ph3 lens) allow for the structural details in the spore to be seen.
2. We usually use the phase contrast mode (100 ms exposure time) to focus on the sample to avoid bleaching the sample. Phase brightness also provides secondary confirmation of the germination status of the spores (Fig. 3a).
3. Typical acquisition parameters: DAPI long-pass filter (excitation filter 365/12, dichroic 395, emission 397 LP), 10-ms exposure time without binning using a Hamamatsu ORCA flash 4.0 v3 CMOS camera, Metamorph Series 7.7 (M7.7) software.
4. Dormant *C. difficile* spores stained with Eu[TTA]$_3$ rapidly bleach under fluorescence imaging, so they appear dimmer compared to germinated spores (Fig. 3a).
5. Dormant and germinated *C. difficile* spores stained with Eu [TTA]$_3$ exhibit staining of the outer layers (i.e., coat and exosporium) but not the core. This contrasts with other bacterial spores, where germination leads to a significant increase in core fluorescence. Since *C. difficile* spores do not undergo this shift in fluorescence during germination, it cannot be used as a correlate of germination. The total fluorescence intensity of the spore serves instead to better distinguish between dormant and germinated *C. difficile* spores.
6. When spores are clumped or aggregated, it is often difficult to distinguish individual spores under phase contrast. In comparison, spores imaged under lanthanide fluorescence are clearly distinguishable as individual spores, even if they are aggregated.
7. Eu[TTA]$_3$ can also be used to stain sporulating rods of *C. difficile,* staining the outer layers. Sporulating rods and spores can be distinguished by cell size and morphology (Fig. 3a).

3.3.3 Confocal Microscopy

1. We use a Leica TCS SP8 inverted gSTED super-resolution microscope with a 100× HCX PL APO Objective (N.A. 1.4, oil objective lens), a 405-nm diode laser (except in the FITC–dextran co-stained samples which require a 488-nm pulsed white light laser), and HyD detectors.
2. Images are acquired using the Leica Application Suite X 1.8 (LASX v1.8) software and deconvolution performed using the in-built lightning mode. For co-staining assays, images are acquired under sequential scanning mode to avoid channel

crosstalk. Cropping and image processing can be performed with ImageJ.

3. Typical acquisition parameters for lanthanide beta-diketone staining are 580–700 nm (Eu complexes), 600–780 nm (Sm[TTA]$_3$), 450–570 nm (Tb[TTA]$_3$), intensity range 6.5% (Eu complexes), 15% (Sm[TTA]$_3$ and Tb[TTA]$_3$), 0.5 airy units, digital zoom 8, and gain 100. For co-staining, acquisition parameters are 580–700 nm (Eu[TTA]$_3$), 500–550 nm (FITC–dextran), 440–530 nm (DAPI), intensity range 6.5% (Eu[TTA]$_3$), 10.7% (FITC–dextran), 0.5% (DAPI), 0.5 airy units, digital zoom 8, and gain 100.
4. Similar to wide-field microscopy, dormant spores stained with Eu[TTA]$_3$ are less fluorescent than germinated spores (Fig. 3b).
5. Co-staining with DAPI and FITC–dextran produces vivid spore images with differentially stained detail. While DAPI stains DNA within the core of germinated spores, DAPI is unable to access the core when the spore is dormant, rendering the core nonfluorescent (Fig. 4a). FITC–dextran provides a contrasting counterstain since its high molecular weight prevents its entry into the outer layers of the spore (Fig. 4b). Finally, Eu[TTA]$_3$ defines the outer boundary of the spore by staining its outer layers.
6. Different lanthanides and beta-diketones can be combined to vary the excitation/emission profiles of the fluorescence. This may prove useful to minimize spectral overlap with other co-staining dyes. Just as with Eu[TTA]$_3$, other lanthanide/beta-diketone combinations produce brighter staining on germinated spores compared to dormant spores (Fig. 5). The spectral emission of the dye is determined by the lanthanide ion. Hence, Eu- and Sm-based dyes fluoresce in the red region while Tb-based dyes fluoresce in the green region. Among the combinations we have tested, Eu[TTA]$_3$ provides the best visual contrast and is recommended as the first choice for staining.

4 Notes

1. We use a Thermo Scientific Sorvall ST 40R for regular centrifugation and a Beckman Avanti J-20 XP for isopycnic centrifugation.
2. The indicated concentrations of lanthanide beta-diketone dyes have been optimized for spore concentrations of 10^7 CFU/mL and above. Lower spore concentrations may require further optimization of dye concentration.

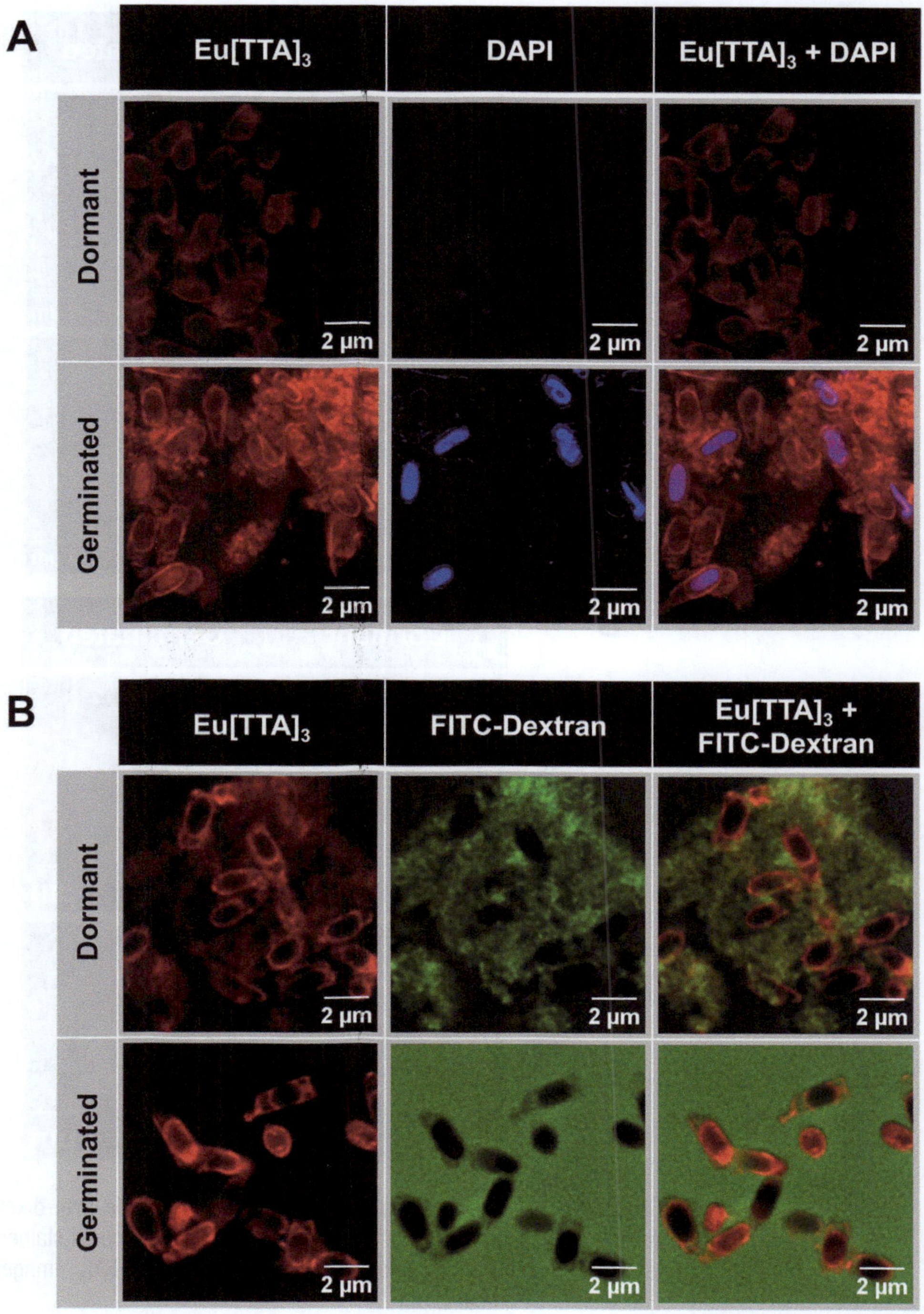

Fig. 4 Co-staining of *C. difficile* spores using Eu[TTA]$_3$ and other dyes. (**a**) Confocal fluorescence images of Eu[TTA]$_3$ and DAPI-stained dormant (top) and germinated (bottom) *C. difficile* spores. Germinated spores exhibit coiled DNA in the spore core (DAPI), while Eu[TTA]$_3$ stains the outer layers of both dormant and germinated spores. (**b**) Confocal fluorescence images of Eu[TTA]$_3$ and FITC–dextran (3–5 kDa)-stained dormant (top) and germinated (bottom) spores. FITC–dextran fails to stain the spores due to its size, while Eu[TTA]$_3$ readily stains the outer layers. Images are pseudocolored. Scale bar 2 μm

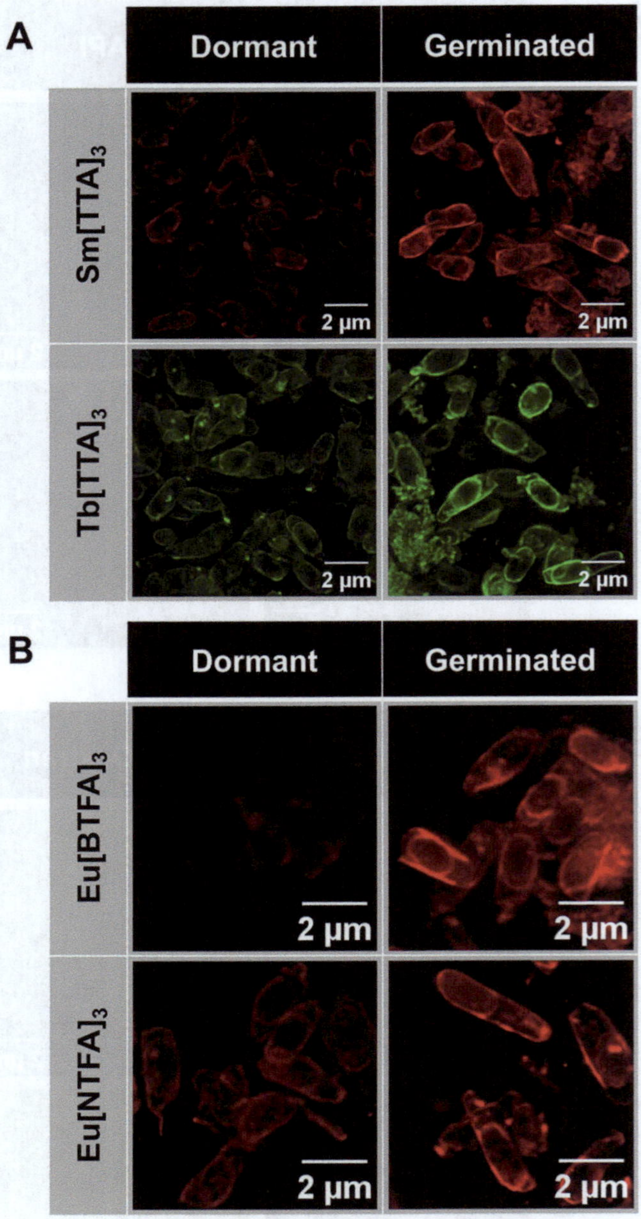

Fig. 5 *C. difficile* spore imaging using alternative lanthanide beta-diketones. Confocal fluorescence images of dormant and germinated spores stained with (**a**) $Sm[TTA]_3$ and $Tb[TTA]_3$ and (**b**) $Eu[BTFA]_3$ and $Eu[NTFA]_3$. Images are pseudocolored. Scale bar 2 μm

3. In general, aliphatic beta-diketones do not fluoresce, and only aromatic beta-diketones function effectively as antenna ligands for lanthanides. This is because aromatic beta-diketones exhibit better photon absorption and higher energy transfer efficiency between the triplet state of the ligand and Eu resonance levels [18].

4. As dye fluorescence can be influenced by environmental factors, absolute quantification of fluorescence signals is not meaningful. Instead, normalized fluorescence changes relative to appropriate controls provide more reliable comparisons.
5. Before using any dye for co-staining with lanthanide beta-diketones, check for chemical interactions and spectral overlap to avoid channel bleed-through effects.
6. Since lanthanide beta-diketone microscopy requires UV excitation, always wear appropriate PPE to avoid UV exposure.

References

1. Bouza E (2012) Consequences of Clostridium difficile infection: understanding the healthcare burden. Clin Microbiol Infect 18(Suppl 6):5–12
2. Zhu D, Sorg JA, Sun X (2018) Clostridioides difficile biology: sporulation, germination, and corresponding therapies for C. difficile infection. Front Cell Infect Microbiol 8:29
3. Shen A (2020) Clostridioides difficile spore formation and germination: new insights and opportunities for intervention. Ann Rev Microbiol 74:545–566
4. Mccormick NG (1965) Kinetics of spore germination. J Bacteriol 89:1180–1185
5. Sundaresan A, Le Ngoc M, Wew MU et al (2023) A design of experiments screen reveals that Clostridium novyi-NT spore germinant sensing is stereoflexible for valine and its analogs. Commun Biol 6:118
6. Hindle AA, Hall EA (1999) Dipicolinic acid (DPA) assay revisited and appraised for spore detection. Analyst 124:1599–1604
7. Sum R, Lim SJM, Sundaresan A et al (2024) Clostridium septicum manifests a bile salt germinant response mediated by Clostridioides difficile csp gene orthologs. Commun Biol 7: 947
8. Kong L, Zhang P, Wang G et al (2011) Characterization of bacterial spore germination using phase-contrast and fluorescence microscopy, Raman spectroscopy and optical tweezers. Nat Protoc 6:625–639
9. Bünzli J-CG, Piguet C (2005) Taking advantage of luminescent lanthanide ions. Chem Soc Rev 34:1048–1077
10. Sundaresan A, Cheong I (2024) Elucidating bacterial spore dynamics through lanthanide-enhanced live imaging. ACS Sens 9:789–798
11. Krick EL, Sorenmo KU, Rankin SC et al (2012) Evaluation of clostridium novyi-NT spores in dogs with naturally occurring tumors. Am J Vet Res 73:112–118
12. Cheong I, Huang X, Bettegowda C et al (2006) A bacterial protein enhances the release and efficacy of liposomal cancer drugs. Science 314:1308–1311
13. Bettegowda C, Huang X, Lin J et al (2006) The genome and transcriptomes of the anti-tumor agent Clostridium novyi-NT. Nat Biotechnol 24:1573–1580
14. Vijay S, Nair RR, Sharan D et al (2023) Percoll discontinuous density gradient centrifugation method for the fractionation of the subpopulations of Mycobacterium smegmatis and Mycobacterium tuberculosis from in vitro cultures. MethodsX 11:102344
15. Chowdhury S, Fried KD, Iwakiri Y et al (2023) Protocol for enrichment, purification, and cytocentrifugation of mouse liver endothelial cells. STAR Protoc 4:102480
16. Edwards AN, McBride SM (2016) Isolating and purifying Clostridium difficile spores. Methods Mol Biol 1476:117–128
17. Yip C, Phan JR, Abel-Santos E (2023) Mechanism of germination inhibition of Clostridioides difficile spores by an aniline substituted cholate derivative (CaPA). J Antibiot (Tokyo) 76:335–345
18. Binnemans K (2005) Chapter 225 - Rare-earth beta-diketonates. In: Gschneidner KA, Bünzli J-CG, Pecharsky VK (eds) Handbook on the physics and chemistry of rare earths. Elsevier, pp 107–272

Chapter 8

Nanoluciferase-Based Reporters for *Clostridioides difficile*

Jeroen Corver, Ana M. Oliveira Paiva, and Wiep Klaas Smits

Abstract

Nanoluciferase (NanoLuc®) is a small, bright reporter enzyme known for its versatility. Detection of nanoluciferase, dependent on the substrate furimazine, is highly sensitive and can be adapted to multiple cellular assays. This chapter describes the use of codon-optimized nanoluciferase as a reporter for promoter activity, to assess protein topology and to investigate protein–protein interactions in *Clostridioides difficile*.

Key words Luciferase, Reporter gene assay, Gene expression

1 Introduction

The use of reporter molecules in molecular microbiology is pivotal for better understanding of fundamental processes in bacteria. For instance, reporters are used for investigation of promoter activity, localization and topology of proteins within a cell, and in vivo protein–protein interactions. The most commonly used reporters are fluorescent proteins (like green fluorescent protein [GFP] and mCherry) and enzymatic reporters (like PhoZ, β-galactosidase, and β-glucuronidase) [1–4]. However, these reporters show one or more limitations, such as a lack of sensitivity, the need for lysis of the bacterial cells to obtain quantitative measurements, and a requirement for oxygen (which limits the use in live anaerobic bacterial cells) [1, 5].

Luciferase proteins are a group of reporters that are characterized by high sensitivity. Reporter measurements are generally fast and in certain cases can be performed in real time [6]. Firefly and Renilla luciferase are relatively large proteins that have been extensively used in eukaryotic biology research, mainly as a means to measure promoter activity. However, the relatively large size, combined with low stability and efficiency restricts the usage of these enzymes [7]. In bacteria, luciferase-based expression analyses are commonly performed with systems derived from the *lux*-operons

Kurt H. Piepenbrink and Jennifer M. Auchtung (eds.), *Clostridioides difficile: Methods and Protocols*, Methods in Molecular Biology, vol. 3046, https://doi.org/10.1007/978-1-0716-5328-9_8,

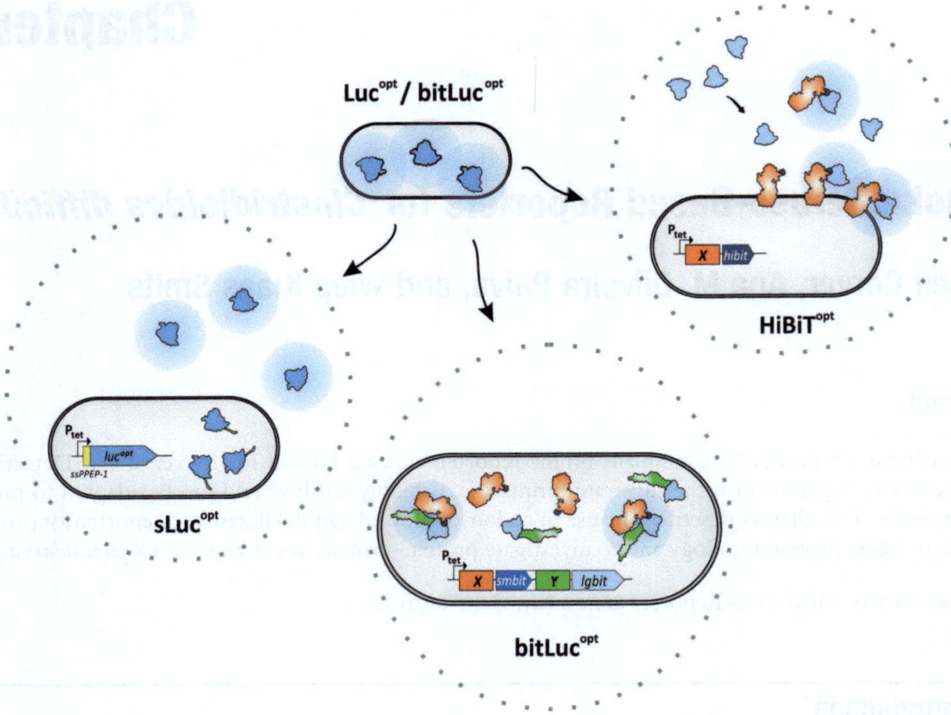

Fig. 1 Schematic representation of the nanoluciferase-derived systems for studies in *C. difficile*. The secreted luciferase reporter system sLuc^{opt} allows for the extracellular detection of the luciferase activity, enabling quick and easy detection for gene expression studies. The Luc^{opt} and bitLuc^{opt} proteins can be used as intracellular reporters. The split luciferase system based on bitLuc^{opt} provides a valuable tool for protein–protein studies in vivo by tagging the proteins of interest (X,Y) with SmBit and LgBit. The HiBiT^{opt} system allows the study of secreted proteins or complexes present at the cell surface and can also be used as a tag for quantitative measurement of protein levels

of bioluminescent bacteria, such as *Vibrio*, *Photobacterium*, and *Photorhabdus* (*Xenorhabdus*) [8]. Yet the fact that these are multi-gene systems limits the application in heterologous bacteria with less-developed genetic systems [6].

Around 13 years ago, a novel luciferase (NanoLuc®) was derived from a deep-sea shrimp luciferase [9]. Since then, this relatively small (19-kD) highly stable enzyme and its novel substrate (furimazine) have been developed as an attractive alternative reporter molecule that compares favorably to other luciferases in terms of brightness [7, 10]. Nanoluc® applications include reporter assays and NanoBRET® and NanoBiT® for investigation of protein–protein interactions [11].

Motivated by these reports, we and others have engineered and used codon-optimized nanoluc-derived constructs for the detection of promoter activity [5, 12–16], localization and topology studies [14, 17], and protein–protein interactions [18–21] in *Clostridioides difficile* (Fig. 1).

2 Materials

2.1 Media and Culture Conditions

We generally use brain–heart infusion medium supplemented with 0.5% yeast extract (BHIY) to culture *C. difficile* on solid and in liquid media.

2.1.1 Bacterial Culture Media

BHIY medium: 37 g/L brain–heart infusion (e.g., Oxoid CM1135B), 5 g/L yeast extract (e.g., Sigma-Aldrich 70161) is used for routine culturing of *C. difficile*. For solid media, this is supplemented with 15 g/L of bacteriological agar (e.g., Thermo Scientific J10906-P2).

To the best of our knowledge, no interactions of medium components under the conditions used for the luciferase assays have been described, and we therefore expect that the choice of medium can be adapted to other commonly used media (including defined media) [5, 22–24] should your experiments require this, without a negative impact on the luminescence measurements.

2.1.2 Culture Conditions

For all plasmid-based systems described here, the engineered plasmids need to be conjugated to the strain of interest according to previously published methods [25–27]. Proper selection should yield the strain of interest.

Transconjugant strains harboring plasmids need to be cultured in or on BHIY supplemented with thiamphenicol (15 μg/mL) to maintain selection of the relevant plasmids. Failure to do so may lead to loss of the plasmid in (a subset) of *C. difficile* cells, despite efficient maintenance of plasmids with a pCD6 replicon (such as those described here) [26, 28], in particular if (over) expression of the gene of interest negatively impacts fitness. Thiamphenicol stock solution: 15 mg/mL in DMSO (1000× stock solution).

2.2 Usage of Nanoluciferase as a Reporter for Promoter Activity

NanoLuc® luciferase catalyzes a highly efficient bioluminescent reaction using the substrate furimazine [7]. This ATP-independent reaction occurs in the presence of molecular oxygen, converting furimazine to furimamide and emitting a bright, stable glow-type luminescence peaking at 460 nm [7].

The luciferase variant sLuc^{opt} is designed for extracellular secretion and emits a luminescent signal detectable in the culture medium, which allows sensitive monitoring of promoter activity without complicated sample processing. This secretion is achieved by fusing the codon-optimized nanoluciferase Luc^{opt} to the signal sequence of PPEP-1, an abundant extracellular metalloprotease naturally secreted by *C. difficile*, directing the export of the reporter protein through the native secretion pathway [5].

We recommend the use of the sLuc^{opt}, as this system has several key advantages. First, due to its efficient secretion and sensitive detection, very small amounts of culture (supernatant) are required for reliable measurements. Second, (sterile) filtered culture supernatant allows assessment of promoter activity at a less-stringent biosafety level. Finally, extracellular production of luciferase minimizes any impact that accumulation of intracellular heterologous nanoluciferase might have on cellular physiology.

Nonquantitative visualization of luciferase using strains carrying promoter-nanoluciferase reporters plated on solid media has also been described [15].

Of note, by coupling $sLuc^{opt}$ expression to the constitutive *fdxA* promoter, nanoluciferase can also effectively be used to monitor bacterial growth by measuring the bioluminescent signal [29], allowing for noninvasive and real-time tracking of cells with a broader dynamic range compared to conventional methods (optical density measurements or fluorescence-based assays).

2.2.1 Plasmids

Several plasmids have been engineered that allow assessment of promoter activity by cloning a promoter sequence in front of a codon-optimized nanoluciferase (Fig. 2).

To study a promoter of interest, the sequence can be cloned into plasmid pAP24 (Addgene #70190) [5] between the KpnI and SacI sites. Alternatively, the pAP24 backbone including $\textit{sluc}^{opt}$ can be amplified by PCR using oligonucleotide primers oWKS-1580 (5'-ATGAGACCAAGTAAAAATTATTAATAGC-3') and oWKS-1582 (5'-GCAAATGCAGGCTTCTTATTTTTATG-3') [15, 16] to insert a sequence of interest using Gibson assembly [30]. This has the added advantage that the native ribosome binding site can be used.

Alternatively, if intracellular nanoluciferase is preferred, Luc^{opt} (identical to $sLuc^{opt}$ but lacking the PPEP-1 signal sequence) [5] or $bitLuc^{opt}$ [18] can be used (Fig. 2). For $bitLuc^{opt}$ both anhydrotetracyclin- (pAF259; Addgene ##105494) [18] and xylose-inducible (pJAK175; Addgene #178601) [31] plasmids have been described. Strains grown under non-inducing conditions, or carrying a construct lacking a promoter sequence (pLKK007; Addgene #245126), can be used as negative controls for luciferase activity.

A well-validated set of plasmids with promoters of various strengths driving the expression of $\textit{bitluc}^{opt}$, functional in *C. difficile* as well as other Clostridia, has been published [32].

2.2.2 Assay Components

To quantify the luciferase secreted by the bacteria, we make use of the Nano-Glo® Luciferase Assay system (Promega N1110). The kit contains two components:

1. Nano-Glo Luciferase Assay Substrate
2. Nano-Glo Luciferase Assay Buffer

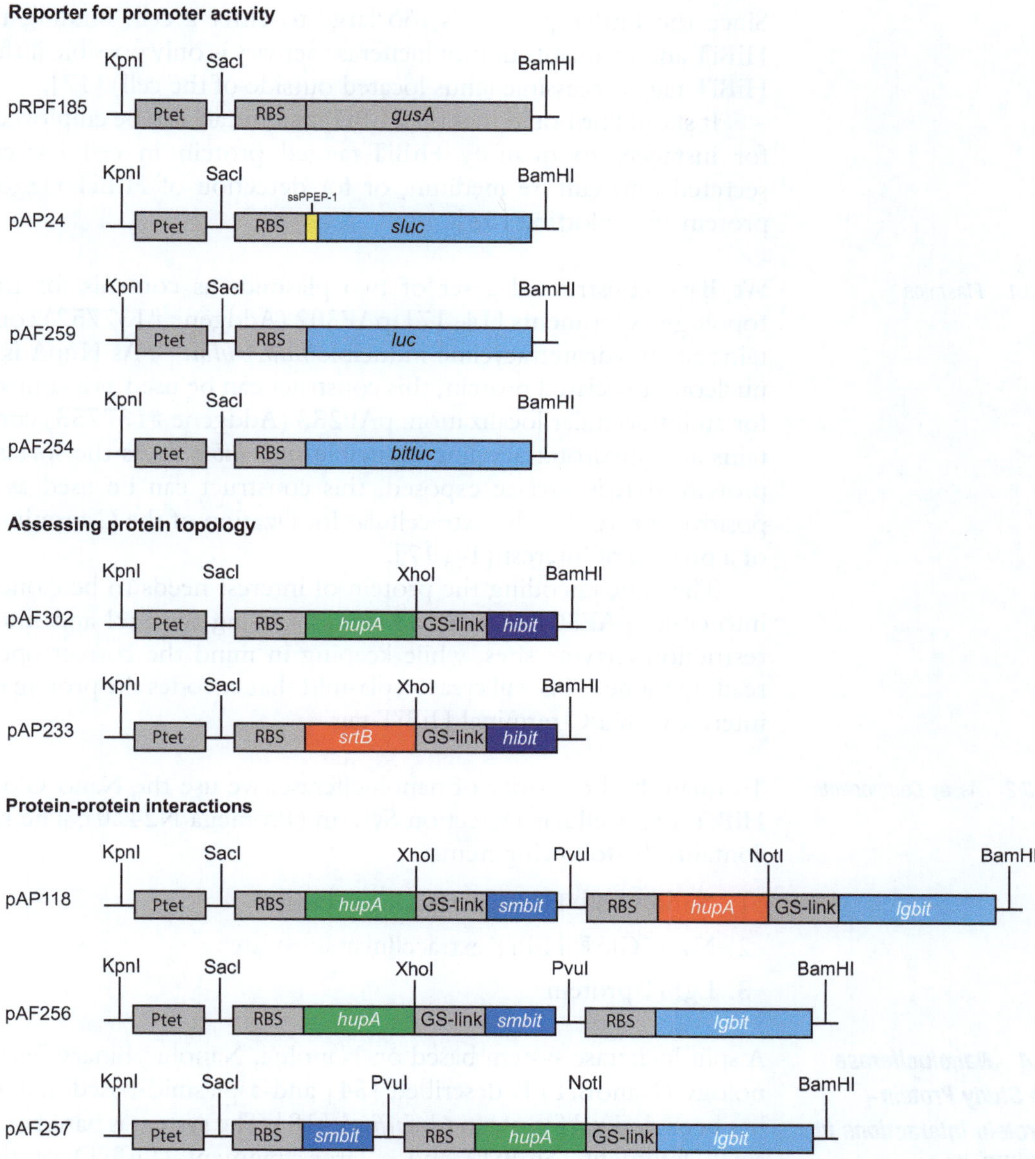

Fig. 2 Overview of plasmid vectors for plasmid-based nanoluciferase tools. Tools are grouped according to the sections of this chapter: reporters for promoter activity, tools for assessing protein topology, and tools for assessing protein–protein interactions

2.3 Nanoluciferase to Determine Topology of Proteins

The NanoLuc® HiBiT extracellular detection system can be used to determine the presence (of part) of a protein in the extracellular space [33]. This system depends on the interaction between the nanoluciferase-derived HiBiT tag (fused to a protein of interest) and extracellularly supplied LgBiT protein. The HiBiT system is engineered for enhanced affinity between the HiBiT tag and LgBiT, which together reconstitute a functional nanoluciferase.

Since the LgBiT protein is too large to enter a cell, binding to HiBiT and reconstitution of luciferase activity is only possible if the HiBiT tag is accessible (thus located outside of the cell) [17].

It should be noted that the HiBiT system can also be employed, for instance, to quantify HiBiT-tagged protein in cell lysates, secreted into culture medium, or for detection of HiBiT-tagged protein after blotting [33].

2.3.1 Plasmids

We have constructed a set of two plasmids as controls for the topology experiments [14, 17]. pAF302 (Addgene #137752) contains an anhydrotetracycline-inducible *hupA-hibit*^{opt}. As HupA is a nucleoid-associated protein, this construct can be used as a control for an intracellular localization. pAP233 (Addgene #137753) contains an anhydrotetracycline-inducible *srtB-hibit*^{opt}. As the sortase protein SrtB is surface exposed, this construct can be used as a positive control for the extracellular localization of the C-terminus of a protein of interest [14, 17].

The gene encoding the protein of interest needs to be cloned into either pAF302 or pAP233 (Fig. 2), using the SacI and XhoI restriction enzyme sites, while keeping in mind the correct open reading frame. This will create a plasmid that encodes the protein of interest with a C-terminal HiBiT-tag.

2.3.2 Assay Components

To quantify the activity of nanoluciferase, we use the Nano-Glo® HiBiT Extracellular Detection System (Promega N2420). The kit contains the following items:

1. Nano-Glo® HiBiT extracellular buffer
2. Nano-Glo® HiBiT extracellular substrate
3. LgBiT protein

2.4 Nanoluciferase to Study Protein–Protein Interactions in C. difficile

A split luciferase system based on Nanoluc, Nanoluc Binary Technology (NanoBiT) is described [34] and a plasmid-based system has been derived for use in *C. difficile* [18]. The system is based on a small fragment (SmBiT) and a large fragment (LgBiT) of the nanoluciferase that have been engineered for reduced self-assembly into a functional nanoluciferase in the absence of fused proteins that show a physical interaction. Bitluc^{opt} is a reconstituted nanoluciferase in which the two engineered subunits have been fused into a single protein, and this protein behaves similar to a nanoluciferase lacking these mutations (Luc^{opt}), demonstrating that the mutations affect only the interaction of SmBiT and LgBit and not nanoluciferase activity per se. Therefore, in our experience, bitLuc^{opt} and Luc^{opt} behave similarly for all intents and purposes.

The published *C. difficile* split luciferase system is plasmid-based and uses an anhydrotetracycline-inducible promoter to drive the co-expression of C-terminal fusions to SmBit and LgBit

[18, 20, 31] and has subsequently also been used in other Firmicutes [35–37].

The split luciferase system has also been used as a diagnostic tool [38], but this application will not be further described here.

2.4.1 Plasmids

Several plasmids are available (Fig. 2) [18].

Plasmid pAF259 (Addgene #105494) encodes *bitluc*^{opt} driven from an anhydrotetracycline-inducible promoter and serves as a positive control for the experiments. Similarly, pAP118 (Addgene #105495) is a positive control: It contains fusion of SmBiT and LgBiT to the nucleoid-associated protein HupA, which homodimerizes in *C. difficile*. Crucial negative controls include pAF256 (Addgene #105496) and pAF257 (Addgene #105497) that harbor the *hupA* gene in frame to either SmBiT or LgBit, respectively, but expressing the other subunit as a non-fusion protein.

Other negative controls have been described (such as constructs expressing only SmBiT and LgBiT, alone or in combination, but offer little added value over the essential controls described above) [18]. pAP118, pAF256, and pAF257 contain restriction sites for replacing *hupA* with gene(s) of interest (Fig. 2): SacI/XhoI (SmBiT-fusion in pAP118 or pAF256) and PvuI/NotI (LgBiT-fusion in pAP118 and pAF257).

2.4.2 Assay Components

Assay components are identical to those described under Subheading 2.2.2.

3 Methods

3.1 Nanoluciferase as a Reporter for Promoter Activity

1. Plate the *C. difficile* strains of interest, containing plasmids with promoter sequences driving the expression of the selected nanoluciferase variant, onto BHIY plates, supplemented with 15 μg/mL thiamphenicol (BHIY-Thia^{15}) (*see* Subheading 2.1). Incubate 24–48 h at 37 °C under anaerobic conditions.
2. Preincubate eight tubes with 5 mL BHIY-Thia^{15} per strain, for at least 4 h at 37 °C, under anaerobic conditions.
3. Inoculate one tube with one colony of the strain of interest and make a ten-fold serial dilution in the remaining seven tubes.
4. Incubate the serial dilution overnight (16 h) at 37 °C under anaerobic conditions, while shaking, and preincubate an Erlenmeyer flask or tube (depending on your plans) containing an appropriate amount of BHIY-Thia^{15} at 37 °C under anaerobic conditions. We generally use 25 mL for 250-mL Erlenmeyer flasks, or 5 mL for glass tubes.

5. Measure the OD_{600} of the highest dilution that shows substantial growth and use that dilution to inoculate the Erlenmeyer or tube at a starting optical density at 600 nm (OD_{600nm}) of 0.05.
6. Follow the growth of the culture by OD_{600nm} at 30–60-min intervals and take samples (100 μL to 1 mL) at time(s) of choice. We usually take samples at least at the log, transition, and stationary phase. Determine also the OD_{600} at the time the samples are taken.
7. Optional: Centrifuge samples for 5 min at 20.000 × g and transfer supernatant to a fresh tube to remove bacterial cells (or filter over a 0.2-μM filter to sterilize culture supernatant).
8. Dilute samples 100-fold in BHIY medium and transfer 3 × 100 ul into a white flat-bottom 96-well plate (technical replicate).
9. Prepare reconstituted substrate solution: Dilute an appropriate amount of Nano-Glo Luciferase substrate 50-fold into the Nano-Glo Luciferase Assay Buffer. Make 20 μL for each well, plus 5% extra to allow for pipetting errors.
10. Add 20 μL reconstituted substrate solution to each well and mix by pipetting up and down. In case of a large number of samples, this step can be carried out with a multichannel pipet.
11. Incubate at room temperature for 5–10 min.
12. Measure luminescence in an appropriate luminometer. We use the Promega Glomax® Explorer Multimode Microplate Reader, but other luminometers or multimode readers may perform well too.
13. For analysis, normalize the relative light units for the number of cells, by dividing the RLU by the OD_{600nm} value. Alternatively, cells can be diluted to a specific optical density prior to measurement, to allow direct comparison. Note: If anhydrotetracycline-inducible expression of the luciferase reporter is used as a control, be sure to include samples that have been induced for >1 h at 200 ng/mL anhydrotetracycline (or 1% xylose) in your measurements.
14. Statistical analyses of differences can be performed with relevant software packages.

3.2 NanoLuc as a Reporter for Protein Topology/Localization

1. Follow **steps 1–5** of Subheading 3.1 for strains carrying plasmids for expression of C-terminal fusions of your protein of interest to HiBiT (*see* Subheading 2.2).
2. Follow the growth of the culture by OD_{600} and induce expression of the gene of interest at OD_{600} approximately 0.5 (log phase) and incubate for 2 h. Timing of expression is dependent on inducible promoter, and gene of interest, and may therefore

have to be optimized. In general, we observe reliable signals >1 h of induction at 200 ng/mL of anhydrotetracycline.

3. Prepare an appropriate volume of assay mix (consisting of LgBiT protein and substrate, diluted 100- and five-fold, respectively, in kit buffer). You need 50 μL per sample, and we recommend to allow 5% extra for pipetting errors.
4. Harvest cells by pelleting 1 mL of bacterial cell culture (1 min 4000 × g) and resuspending the pellet in phosphate-buffered saline.
5. Optional: Prepare a total bacterial cell lysate from part of the suspension of point 4 and sonicating the cells, or enzymatically lysing them (*see* **Notes 1–6**). This step makes intracellularly located HiBiT tags available to the LgBiT protein and therefore acts as a control for proper expression of the construct. Similarly, total cell lysates can be separated on SDS-PAGE and probed after blotting with LgBiT solution to verify the expression of HiBiT-tagged proteins.
6. Take a 3× 50 μL sample of the bacterial cell suspension (technical replicate) and mix this 50 μL with 50 μL of the assay mix in a white flat-bottom 96-well plate. If relevant, do the same for the cell lysate.
7. Incubate at room temperature for 5–10 min.
8. Measure luminescence in an appropriate luminometer. We use the Glomax® Explorer Multimode Microplate Reader (Promega).
9. For analysis, normalize the relative light units (RLUs) to the OD_{600} value, as a proxy for the total number of cells.
10. Statistical analyses of differences can be performed with relevant software packages.

3.3 NanoLuc as a Reporter for Protein–Protein Interaction

1. Follow **steps 1–5** of Subheading 3.1 for strains carrying plasmids for expression of C-terminal fusions of your protein of interest to SmBiT, LgBiT or both (*see* Subheading 2.3).
2. Follow the growth of the culture by OD_{600} and induce expression of the gene(s) of interest at OD_{600} approximately 0.5 (log phase) and incubate for 2 h. Timing of expression is dependent on inducible promoter, and gene of interest, and may therefore have to be optimized. In general, we observe reliable signals >1 h of induction at 200 ng/mL of anhydrotetracycline.
3. Prepare reconstituted substrate solution: Dilute an appropriate amount of Nano-Glo Luciferase substrate 50-fold into the Nano-Glo Luciferase Assay Buffer. Make 20 μL for each well, plus 5% extra to allow for pipetting errors.

4. Transfer 3 × 100 uL into a white flat-bottom 96-well plate (technical replicate).
5. Add 20 μL reconstituted substrate solution to each well and mix by pipetting up and down. In case of a large number of samples, this step can be carried out with a multichannel pipet.
6. Incubate at room temperature for 5–10 min.
7. Measure luminescence in an appropriate luminometer. We use the Glomax® Explorer Multimode Microplate Reader (Promega).
8. For analysis, normalize the relative light units (RLU) to the OD_{600} value, as a proxy for the total number of cells.
9. Statistical analyses of differences can be performed with relevant software packages.

4 Notes

Luciferase measurements, as a result of their high sensitivity, are highly susceptible to experimental variation. In order to obtain reliable measurements, we offer the following considerations:

1. *Instrumentation*. Reliability of measurements across microtiter plates should be assessed under the conditions used for measurements. If measurements are performed under aerobic conditions in the presence of bacterial cells, it should be assessed if no major stress responses are initiated that affect the expression of the luciferase reporters.
2. *Plate layout*. To minimize luminescence bleed-through or cross-talk in a plate-based luminescence assay, an effective approach is to add empty wells or buffer-only control wells between the sample wells. This helps to achieve more accurate and reliable measurements by dampening the diffusion of luminescent light into neighboring wells and reducing the interference caused by bleed-through in particular in the case of wells containing high signals adjacent to wells with low signals.
3. *Replicates*. We recommend the inclusions of both technical and biological replicates ($n \geq 3$).
4. *Inducible expression*. In exceptional cases, expression from anhydrotetracycline-inducible promoters may interfere with the assays (due to leakiness, high-level expression); if lower expression (by lowering the level of inducer) does not remedy the problem, consider using alternative inducible promoters with better compatible characteristics (e.g., Pxyl) [31, 39], or using native promoters. The anhydrotetracycline-inducible promoter can be exchanged in the plasmids using a KpnI/SacI cloning strategy (Fig. 2). For inducible expression

constructs, we recommend splitting the relevant cultures prior to induction, to ensure directly comparable starting materials.

5. *Plasmids.* The systems described above are based on multicopy plasmids. However, as noted, it requires the continuous selection for the plasmid, which in rare instances could affect luciferase read-outs. If this is the case, the relevant features of the systems described can be cloned into allelic exchange vectors [40–44] and introduced onto the *C. difficile* chromosome. This has been successfully done, for instance, in *Streptococcus pneumoniae* for SmBiT and LgBiT [45].
6. *Enzymatic lysis.* Efficient lysis of most if not all *C. difficile* strain can be achieved using the catalytic domain of endolysin CD27L that can be recombinantly produced and purified in *E. coli* using plasmid pHAS042 [46, 47].

References

1. Oliveira Paiva AM, Friggen AH, Douwes R, Wittekoek B, Smits WK (2022) Practical observations on the use of fluorescent reporter systems in Clostridioides difficile. Antonie Van Leeuwenhoek 115(2):297–323
2. Edwards AN, Pascual RA, Childress KO, Nawrocki KL, Woods EC, McBride SM (2015) An alkaline phosphatase reporter for use in Clostridium difficile. Anaerobe 32:98–104
3. Ransom EM, Ellermeier CD, Weiss DS (2015) Use of mCherry red fluorescent protein for studies of protein localization and gene expression in Clostridium difficile. Appl Environ Microbiol 81(5):1652–1660
4. Mani N, Dupuy B (2001) Regulation of toxin synthesis in Clostridium difficile by an alternative RNA polymerase sigma factor. Proc Natl Acad Sci USA 98(10):5844–5849
5. Oliveira Paiva AM, Friggen AH, Hossein-Javaheri S, Smits WK (2016) The signal sequence of the abundant extracellular metalloprotease PPEP-1 can be used to secrete synthetic reporter proteins in Clostridium difficile. ACS Synth Biol 5(12):1376–1382
6. Jiang T, Bai X, Li M (2024) Advances in the development of bacterial bioluminescence imaging. Annu Rev Anal Chem (Palo Alto, Calif) 17(1):265–288
7. England CG, Ehlerding EB, Cai W (2016) NanoLuc: a small luciferase is brightening up the field of bioluminescence. Bioconjug Chem 27(5):1175–1187
8. Close D, Xu T, Smartt A, Rogers A, Crossley R, Price S et al (2012) The evolution of the bacterial luciferase gene cassette (lux) as a real-time bioreporter. Sensors (Basel) 12(1): 732–752
9. Hall MP, Unch J, Binkowski BF, Valley MP, Butler BL, Wood MG et al (2012) Engineered luciferase reporter from a deep sea shrimp utilizing a novel imidazopyrazinone substrate. ACS Chem Biol 7(11):1848–1857
10. Loh JM, Proft T (2014) Comparison of firefly luciferase and NanoLuc luciferase for biophotonic labeling of group A Streptococcus. Biotechnol Lett 36(4):829–834
11. Dale NC, Johnstone EKM, White CW, Pfleger KDG (2019) NanoBRET: the bright future of proximity-based assays. Front Bioeng Biotechnol 7:56
12. Pannullo AG, Guan Z, Goldfine H, Ellermeier CD (2023) HexSDF is required for synthesis of a novel glycolipid that mediates daptomycin and bacitracin resistance in C. difficile. mBio 14(2):e0339722
13. Pannullo AG, Zbylicki BR, Ellermeier CD (2023) Identification of DraRS in Clostridioides difficile, a two-component regulatory system that responds to Lipid II-interacting antibiotics. J Bacteriol 205(10):e0016423
14. Corver J, Claushuis B, Shamorkina TM, de Ru AH, van Leeuwen MM, Hensbergen PJ et al (2024) Proteolytic activity of surface-exposed HtrA determines its expression level and is needed to survive acidic conditions in Clostridioides difficile. Mol Microbiol 122(4): 413–428
15. Boekhoud IM, Michel AM, Corver J, Jahn D, Smits WK (2020) Redefining the Clostridioides difficile sigma(B) Regulon: sigma

(B) activates genes involved in detoxifying radicals that can result from the exposure to antimicrobials and hydrogen peroxide. mSphere. 5(5)

16. van Eijk E, Boekhoud IM, Kuijper EJ, Bos-Sanders I, Wright G, Smits WK (2019) Genome location dictates the transcriptional response to PolC inhibition in Clostridium difficile. Antimicrob Agents Chemother 63(2)
17. Oliveira Paiva AM, de Jong L, Friggen AH, Smits WK, Corver J (2020) The C-terminal domain of Clostridioides difficile TcdC is exposed on the bacterial cell surface. J Bacteriol 202(22)
18. Oliveira Paiva AM, Friggen AH, Qin L, Douwes R, Dame RT, Smits WK (2019) The bacterial chromatin protein HupA can remodel DNA and associates with the nucleoid in Clostridium difficile. J Mol Biol 431(4):653–672
19. DiCandia MA, Edwards AN, Jones JB, Swaim GL, Mills BD, McBride SM (2022) Identification of functional Spo0A residues critical for sporulation in Clostridioides difficile. J Mol Biol 434(13):167641
20. Nerber HN, Baloh M, Brehm JN, Sorg JA (2024) The small acid-soluble proteins of Clostridioides difficile regulate sporulation in a SpoIVB2-dependent manner. PLoS Pathog 20(8):e1012507
21. Baloh M, Sorg JA (2021) Clostridioides difficile SpoVAD and SpoVAE interact and are required for dipicolinic acid uptake into spores. J Bacteriol 203(21):e0039421
22. Karasawa T, Ikoma S, Yamakawa K, Nakamura S (1995) A defined growth medium for Clostridium difficile. Microbiology (Reading). 141 (Pt 2):371–375
23. Cartman ST, Minton NP (2010) A mariner-based transposon system for in vivo random mutagenesis of Clostridium difficile. Appl Environ Microbiol 76(4):1103–1109
24. Karlsson S, Burman LG, Akerlund T (1999) Suppression of toxin production in Clostridium difficile VPI 10463 by amino acids. Microbiology (Reading). 145(Pt 7):1683–1693
25. Heap JT, Pennington OJ, Cartman ST, Carter GP, Minton NP (2007) The ClosTron: a universal gene knock-out system for the genus clostridium. J Microbiol Methods 70(3): 452–464
26. Purdy D, O'Keeffe TA, Elmore M, Herbert M, McLeod A, Bokori-Brown M et al (2002) Conjugative transfer of clostridial shuttle vectors from Escherichia coli to Clostridium difficile through circumvention of the restriction barrier. Mol Microbiol 46(2):439–452
27. Kirk JA, Fagan RP (2016) Heat shock increases conjugation efficiency in Clostridium difficile. Anaerobe 42:1–5
28. Smits WK, Roseboom AM, Corver J (2022) Plasmids of Clostridioides difficile. Curr Opin Microbiol 65:87–94
29. Schumacher J, Muller P, Sulzer J, Faber F, Molitor B, Maier L (2025) Proton-pump inhibitors increase C. difficile infection risk by altering pH rather than by affecting the gut microbiome based on a bioreactor model. Gut Microbes 17(1):2519697
30. Gibson DG, Young L, Chuang RY, Venter JC, Hutchison CA 3rd, Smith HO (2009) Enzymatic assembly of DNA molecules up to several hundred kilobases. Nat Methods 6(5): 343–345
31. Phothichaisri W, Phetruen T, Chankhamhaengdecha S, Janvilisri T, Ounjai P, Fagan RP et al (2024) Unraveling physical interactions of Clostridioides difficile with phage and phage-derived proteins using in vitro and whole-cell assays. Methods Mol Biol 2738:245–262
32. Fisher H, de Lussy Kubisa L, Jakhmola A, Walker E, Kirk JA, Oatley P et al (2025) A set of genetic tools for use in Clostridioides difficile and related species. Microbiology (Reading)
33. HiBiT Protein Tagging System: Promega Corporation; Available from: https://nld.promega.com/resources/technologies/hibit-protein-tagging-system/
34. Dixon AS, Schwinn MK, Hall MP, Zimmerman K, Otto P, Lubben TH et al (2016) NanoLuc complementation reporter optimized for accurate measurement of protein interactions in cells. ACS Chem Biol 11(2): 400–408
35. Gallay C, Sanselicio S, Anderson ME, Soh YM, Liu X, Stamsas GA et al (2021) CcrZ is a pneumococcal spatiotemporal cell cycle regulator that interacts with FtsZ and controls DNA replication by modulating the activity of DnaA. Nat Microbiol 6(9):1175–1187
36. Barbuti MD, Lambert E, Myrbraten IS, Ducret A, Stamsas GA, Wilhelm L et al (2024) The function of CozE proteins is linked to lipoteichoic acid biosynthesis in Staphylococcus aureus. MBio 15(6):e0115724
37. Wood A, Irving SE, Bennison DJ, Corrigan RM (2019) The (p)ppGpp-binding GTPase era promotes rRNA processing and cold adaptation in Staphylococcus aureus. PLoS Genet 15(8):e1008346
38. Adamson H, Ajayi MO, Gilroy KE, McPherson MJ, Tomlinson DC, Jeuken LJC (2022) Rapid

quantification of C. difficile glutamate dehydrogenase and toxin B (TcdB) with a NanoBiT split-luciferase assay. Anal Chem 94(23): 8156–8163

39. Muh U, Pannullo AG, Weiss DS, Ellermeier CD (2019) A xylose-inducible expression system and a CRISPR interference plasmid for targeted knockdown of gene expression in Clostridioides difficile. J Bacteriol 201(14)
40. Cartman ST, Kelly ML, Heeg D, Heap JT, Minton NP (2012) Precise manipulation of the Clostridium difficile chromosome reveals a lack of association between the tcdC genotype and toxin production. Appl Environ Microbiol 78(13):4683–4690
41. Ng YK, Ehsaan M, Philip S, Collery MM, Janoir C, Collignon A et al (2013) Expanding the repertoire of gene tools for precise manipulation of the Clostridium difficile genome: allelic exchange using pyrE alleles. PLoS One 8(2):e56051
42. Maikova A, Kreis V, Boutserin A, Severinov K, Soutourina O (2019) Using an endogenous CRISPR-Cas System for genome editing in the human pathogen Clostridium difficile. Appl Environ Microbiol 85(20)
43. Peltier J, Hamiot A, Garneau JR, Boudry P, Maikova A, Hajnsdorf E et al (2020) Type I toxin-antitoxin systems contribute to the maintenance of mobile genetic elements in Clostridioides difficile. Commun Biol 3(1):718
44. McAllister KN, Bouillaut L, Kahn JN, Self WT, Sorg JA (2017) Using CRISPR-Cas9-mediated genome editing to generate C. difficile mutants defective in selenoproteins synthesis. Sci Rep 7(1):14672
45. Rengifo-Gonzalez M, Mazzuoli MV, Janssen AB, Rueff AS, Burnier J, Liu X et al (2025) Make-or-break prime editing for genome engineering in Streptococcus pneumoniae. Nat Commun 16(1):3796
46. Mayer MJ, Garefalaki V, Spoerl R, Narbad A, Meijers R (2011) Structure-based modification of a Clostridium difficile-targeting endolysin affects activity and host range. J Bacteriol 193(19):5477–5486
47. Peltier J, Shaw HA, Couchman EC, Dawson LF, Yu L, Choudhary JS et al (2015) Cyclic diGMP regulates production of sortase substrates of Clostridium difficile and their surface exposure through ZmpI protease-mediated cleavage. J Biol Chem 290(40):24453–24469

Chapter 9

Working with *C. difficile* in a Hypoxic Chamber

Léo C. Caulat, Aurélie Lotoux, Isabelle Martin-Verstraete, and Claire Morvan

Abstract

Clostridioides difficile is an enteropathogen responsible for diarrhea under conditions leading to a dysbiosis of the gut microbiota, in particular after an antibiotic treatment. It is a spore-forming obligate anaerobe, which is, as such, usually cultivated in anaerobiosis in laboratories. Yet the gastrointestinal tract (GIT), its natural niche, is not completely deprived of oxygen (O_2) and the O_2 tension in the gut even rises under conditions of dysbiosis. *C. difficile* is able to tolerate the low O_2 tensions it encounters in the GIT. *C. difficile* spores germinate in the small intestine at 4% O_2 and the bacteria are found close to epithelial cells where the O_2 tension is between 1% and 2% O_2. To understand how this anaerobe deals with the oxidative stress found in its environment and to investigate its physiology in atmospheric conditions closer to the O_2 exposure encountered in the GIT, experiments can now be performed in hypoxia chambers in which the O_2 tensions can be modulated. This chapter details how to investigate *C. difficile* biology in hypoxia.

Key words *Clostridioides difficile*, O_2 gradient, Physiological O_2 tensions, Sampling, Hypoxia

1 Introduction

Even though *Clostridioides difficile* is a strict anaerobe, it faces several sources of oxidative stress during its infection cycle and in particular exposure to various O_2 tensions in the gastrointestinal tract (GIT) or outside the host. In the environment, vegetative cells and spores of *C. difficile* face atmospheric O_2 tension (21%). Depending on the strains, vegetative cells survive in air between a few minutes and a few hours at most [1, 2], while spores resist for years [3]. Spores are dormant cells with restricted permeability and a very reduced metabolism that cannot be poisoned by O_2 and prevents endogenous ROS production, explaining why they are the dissemination form of the bacterium.

The GIT is also not a strictly anaerobic environment. There are two gradients of O_2 in the GIT [4–6]. A first gradient of O_2 is longitudinal, decreasing from 7.5% in the stomach to 4% in the

Kurt H. Piepenbrink and Jennifer M. Auchtung (eds.), *Clostridioides difficile: Methods and Protocols*, Methods in Molecular Biology, vol. 3046, https://doi.org/10.1007/978-1-0716-5328-9_9,

small intestine where the spores germinate and to around 0.1–0.4% in the colon where the bacterium multiplies and colonizes the colon. A second lateral increasing gradient exists along the entirety of the GIT increasing from the lumen of the GIT toward the epithelium. In the colon, this gradient increases from 0.1% to 0.4% in the lumen to around 1–2% in the mucus layer where *C. difficile* localizes and up to 5% in the tissues. These low, but still toxic, O_2 tensions constitute a major source of stress for anaerobes and require adaptive strategies [7]. Moreover, *C. difficile* infections are favored by intestinal microbiota dysbiosis, e.g., following antibiotic treatments. Under dysbiotic conditions, the O_2 tensions encountered in the gut rise. This is due to a change in the metabolism of the epithelial cells that switch from O_2-consumming butyrate β-oxidation toward glucose fermentation when butyrate-producing bacteria are depleted [8].

C. difficile is usually grown in an anaerobic atmosphere, either in anaerobic chambers or in anaerobic jars, where the anaerobic atmosphere is maintained by the use of a precise mix of N_2, CO_2, and H_2 or by using Gaspacks. It has been previously shown that exposure to different O_2 tensions changes drastically the transcriptome of *C. difficile* [9–12]. Plus, we showed that long-term adaptation to O_2 affects key enzymes involved in O_2 and reactive oxygen species detoxication but also various key steps of the *C. difficile* lifecycle [1, 2, 13]. Studying the response of *C. difficile* to the low O_2 tensions found in the gut is thus important to better understand the physiopathology of this bacterium and performing experiments in the presence of physiological O_2 tensions is a way to better recapitulate the conditions encountered by the bacterium during the infection.

2 Materials

2.1 Hypoxic Chamber

Hypoxic chambers, like the Bugbox M from Baker-Ruskinn we used, allow the precise control of the atmospheric composition to which cells are exposed. As in anaerobic cabinets, they can be either thermostated or contain a thermostated incubator for optimal bacterial growth. We chose 37 °C for *C. difficile* cultures. The humidity of the chamber should be controlled carefully, as a dry atmosphere would lead to a drought stress and a loss of media, while a too humid atmosphere could facilitate contaminations. The expected humidity level is between 50% and 65%.

The atmosphere composition we used in the hypoxic chamber is 5% CO_2, x% O_2, and (95-x)% N_2. Tensions ranging from 0.1% to 0.4% O_2 were chosen to mimic the low O_2 tensions found in the lumen of the colon. The lowest tension achievable is often 0.1% and O_2 values are often incremented by 0.1%. Thus, while the O_2 sensor of the chamber might indicate that the set O_2 tension is 0.1%, it could range from 0.01% to 0.19% which is a huge difference as a tiny decrease of O_2 tension would lead to anaerobiosis. This could influence the results of experiments and could induce reproducibility issues. So, 0.4% will be preferred to 0.1% O_2. Tensions between 1% and 2% will be chosen to grow cells in the presence of O_2 tensions similar to those found in the vicinity of the epithelial cells. To work at the O_2 tensions found in the upper intestine where *C. difficile* spores germinate, tensions around 4–5% will be selected. We rather recommend 4% as we observed that beyond, the reproducibility of the result is impaired.

We also recommend, when possible, to perform biological replicates on the same day to have cells exposed to the exact same atmosphere as there could be slight fluctuations over time and to perform the experiments at three different moments to exclude any effect of these tiny variations.

2.2 Growth Medium

To perform growth in hypoxia, we recommend rich media, like tryptone yeast (TY). The medium should be supplemented with taurocholate (Tau) at 0.05% (w/v) when looking at vegetative cells, to avoid spore formation.

2.3 Solid Growth Conditions

When growing cells on agar plates, plates should be calibrated in volume to increase the reproducibility and avoid any thinning of the plates due to some dryness of the cabinet atmosphere. We recommend, as an example, to calibrate round 9.4-cm- diameter (e.g., Greiner Bio-One 633,180) Petri dishes with 25 mL of agar medium and 10 by 10 cm square plates (e.g., SIMPORT D210–16) with 28 mL. We recommend the use of square plates with a grid at the bottom to easily spot bacterial cultures in each square of the grid so that they do not mix (5 μL/drop).

In all cases, plates should be dry before spotting so that bacterial culture deposited dries quickly.

2.4 Killing Buffer for Sampling on Agar Medium

The killing buffer is made with 20 mM Tris–HCl 7.5, 5 mM $MgCl_2$, and 20 mM NaN_3. It should be cold and either made extemporaneously or stored at 4 °C before use [14].

2.5 Liquid Growth Medium

Liquid cultures in anaerobic cabinets are classically performed in a few milliliters of medium in plastic tubes of 15- or 50-mL capacity without agitation. When performing static cultures in such tubes in hypoxia, a clear gradient of O_2 forms in the tubes. The bottom of the tubes are murky while the top is clear (Fig. 1), suggesting that the bacteria are preferably growing at the bottom of the tube where the O_2 tension is the lowest. A way to solve this problem can be to agitate gently the culture to favor O_2 diffusion in the medium. However, anaerobic bacteria are classically not grown with agitation.

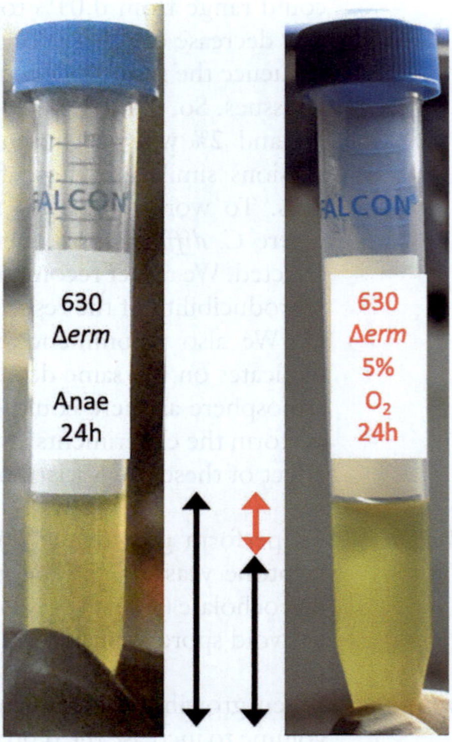

Fig. 1 Establishment of anaerobic condition in tubes cultured in hypoxia. *C. difficile* 630Δ*erm* was grown in liquid TY ON in anaerobiosis and then diluted 1/50 in 5 mL TY Tau in 15-mL tubes. One tube was incubated in anaerobiosis (black writting) while the other was incubated in a hypoxic chamber in the presence of 5% O_2 (red writing) statically for 24 h. Pictures show a zone of less growth at the surface of the medium exposed to hypoxia (red arrow). We hypothetize the medium is anaerobic under this zone (black arrows)

In order to avoid the establishment of an O_2 gradient in the liquid cultures in hypoxia, the surface/volume ratio of the cultures needs to be maximized. We chose to perform the cultures in 6-well plates with 2 mL of culture per well. We found this maximized the surface/volume ratio with enough volume to avoid important loss because of the drought at 37 °C and let enough material for subsequent experiments. If 2 mL is not enough for the desired experimental protocol, the culture volume needed can be split in different wells. Up to 12 mL can be cultured on a 6-well plate. Alternatively, tissue culture flasks can be used with a minimal volume necessary to cover the entire surface of the flask (e.g., 5 mL in 25 cm^2 flasks).

3 Methods

3.1 Preparation of the Bacteria in Anaerobiosis

1. An overnight culture (O/N) is prepared in 5 mL of TY in a 15-mL plastic tube by inoculating a colony grown on agar plate. The O/N culture is let to grow 16 h in anaerobiosis at 37 °C.
2. The O/N culture is subsequently diluted 1:50 in 5 mL of TY to start an overday (O/D) culture.
3. The O/D culture is let to grow at 37 °C for around 3 h to reach mid-exponential phase.

3.2 Assaying Growth and Survival Phenotypes in the Presence of O_2

1. The OD_{600nm} of the culture is then measured in the anaerobic cabinet.
2. An inoculum of 1 mL at an OD of 0.5 is prepared (if the O/D culture has a lower $OD_{600nm} < 0.5$; *see* **Note 1**).
3. This inoculum is serially diluted in TY Tau by 10, up until 10^{-5} in a 96-well plate.
4. Then 5-μL drops of each dilution are spotted on two TY Tau agar plates.
5. Once the spots are dried, one plate is left to incubate for 24 h at 37 °C in anaerobiosis, as a growth control. It will also ensure that the growth of all strains tested is similar in anaerobiosis.
6. The other plates are retrieved from the anaerobic cabinet and incubated at the desired O_2 tension for the desired time at 37 ° C (*see* **Note 2**).
7. If necessary, for instance, for a cell survival assay, plates can be subsequently reincubated for 24 h at 37 °C in anaerobiosis (*see* **Note 3**).
8. CFU count is then made and bacterial concentrations can be assessed. Survival and/or growth in the desired conditions can be compared via normalization by CFU counts in anaerobiosis (*see* **Note 4**).

3.3 Growth on Agar Plate for Sampling

1. A mid-exponential grown culture of 200 μL (cf. 3.1) is plated on two TY agar plates using a spreader (*see* **Note 5**).
2. The plates are then incubated for the desired time at 37 °C, one in anaerobiosis and the other in hypoxia.
3. At the desired sampling times, check that bacteria have grown (either as a lawn or as isolated colonies). To sample the bacteria, resuspend them in 3 mL of killing buffer using a rake.
4. Sample the remaining liquid.

3.4 Growth in Liquid Culture for Sampling

1. The OD_{600nm} of the anaerobic O/D culture is measured (cf. 3.1).
2. Cultures in fresh rich medium are prepared at a starting OD_{600nm} of 0.05.
3. Then 2 mL of this culture are distributed in wells of two 6-well plates with 2 mL per well (*see* **Note 5**).
4. One plate is incubated in anaerobiosis as a control.
5. The second plate is incubated at the selected O_2 tension for the desired time.
6. At the desired sampling time, contents of the wells are mixed by pipetting and sampled.

3.5 Examples of Use of the Samples

1. *Growth monitoring*. The OD_{600nm} of the liquid cultures or the resuspended agar cultures can be measured.
2. *Survival assays*. Plating serial dilutions of liquid cultures on rich medium plates in anaerobiosis and numbering the CFU/mL will give access to survival rates after exposure to hypoxia.
3. *RNA extraction*. Samples can be quickly spun for 30 s at max speed right after sampling, and the pellet flash-frozen in liquid nitrogen. These pellets can then be stored at −80 °C before use.
4. *Other cellular component extraction*, e.g., proteins. Samples can be spun for 30 s at max speed. The pellets can be washed once in 1X PBS and stored at −20 °C prior to analysis.
5. *Spent medium study*. The supernatant of liquid cultures can be kept in a separate tube to measure extracellular concentration of a secreted molecule or for cytotoxicity assay via lactate deshydrogenase dosage [15], for instance.
6. *Fluorescent microscopy*: The same staining protocols can be applied in the presence of O_2 or in anaerobiosis. Cells are sampled, washed once in PBS, fixed with 200 μL of 4% paraformaldehyde for 30 min, and washed thrice in PBS. Staining can be performed before or after fixation depending on the dye used and if it needs O_2 or not. Samples can then be imaged immediately or dry pellets can be stored at 4 °C.

4 Notes

1. Culture with an $OD_{600nm} < 0.5$ needs to be concentrated by centrifugation before being resuspended in 1 mL of fresh media. The calculation is $V_{\text{sample}} = \frac{OD_{\text{goal}} \times V_{\text{goal}}}{OD_{\text{sample}}}$ with V_{sample} the volume to sample in mL, OD_{goal} the desired OD (0.5), V_{goal} the desired volume (1 mL), and OD_{sample} the measured OD_{600nm}. A centrifugation of 2 min at 13,000 rpm is enough to pellet the cells.

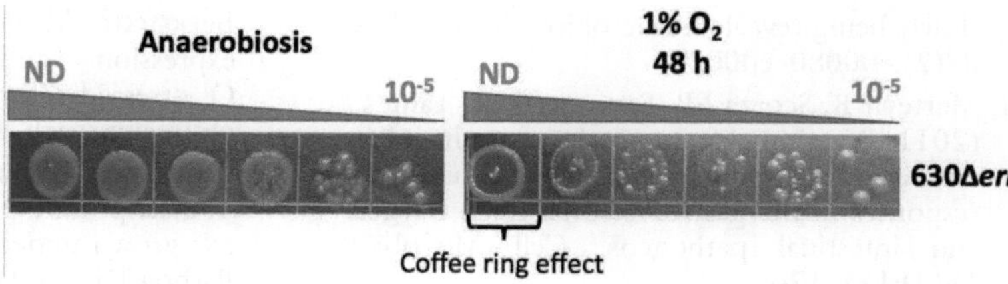

Fig. 2 A coffee-ring shape appears from bacterial drops after growth in the presence of O_2. *C. difficile* was grown in liquid TY for 3 h. An inoculum of OD_{600nm} 0.5 was then prepared and serially diluted by 10 up to 10^{-5} in TY. Then 5 μL of each dilution were plated on calibrated TY Tau agar plates. One plate was incubated in anaerobiosis while the other was incubated in a hypoxic chamber in the presence of 1% O_2 for 48 h. Pictures are representative and show a coffee-ring effect for the plate exposed to hypoxia especially for the non-diluted drop

2. To assay growth at low and intermediate O_2 tensions (up to 1% O_2), maximum growth should be achieved within 48 h in hypoxia. For growth and survival assay at O_2 tensions ranging from 2% to 5%, we recommend times of incubations between 8 and 24 h in hypoxia before 24 h of incubation in anaerobiosis.
3. For tolerance to low and intermediate O_2 tensions (up to 1% O_2), for incubation times shorter than 48 h, we advise to reincubate cells in anaerobiosis to allow growth of surviving cells. To measure survival to O_2 tensions ranging from 2% to 5%, plates need to be further incubated in anaerobiosis for growth of the cells that survived.
4. A coffee-ring effect (Fig. 2) might be observed on colonies grown in hypoxia or air. This is a phenotype often seen [16].
5. More plates or more wells can be prepared from the same culture to increase sample quantity. The samples recovered from those different plates or wells can be pooled.

Acknowledgements

The authors acknowledge the technical support of Emilie Bertrand and Elena Capuzzo as well as the financial support of the ANR Difox (ANR-22-CE15-0026-01).

References

1. Caulat LC, Lotoux A, Martins MC, Kint N, Anjou C, Teixeira M et al (2024) Physiological role and complex regulation of O_2 -reducing enzymes in the obligate anaerobe *Clostridioides difficile*. MBio:e01591–e01524
2. Lotoux A, Caulat L, Alves CM, Alves Feliciano C, Morvan C, Folgosa F et al (2025) Defense arsenal of the strict anaerobe *Clostridioides difficile* against reactive oxygen species encountered during its infection cycle. mBio:e0375324
3. Setlow P, Christie G (2023) New thoughts on an old topic: secrets of bacterial spore resistance

slowly being revealed. Microbiol Mol Biol Rev 87(2):e00080–e00022

4. Marteyn B, Scorza FB, Sansonetti PJ, Tang C (2011) Breathing life into pathogens: the influence of oxygen on bacterial virulence and host responses in the gastrointestinal tract: oxygen and intestinal pathogens. Cell Microbiol 13(2):171–176
5. Albenberg L, Esipova TV, Judge CP, Bittinger K, Chen J, Laughlin A et al (2014) Correlation between intraluminal oxygen gradient and radial partitioning of intestinal microbiota. Gastroenterology 147(5):1055–1063.e8
6. Keeley TP, Mann GE (2019) Defining physiological normoxia for improved translation of cell physiology to animal models and humans. Physiol Rev 99(1):161–234
7. Lu Z, Imlay JA (2021) When anaerobes encounter oxygen: mechanisms of oxygen toxicity, tolerance and defence. Nat Rev Microbiol 19(12):774–785
8. Rivera-Chávez F, Lopez CA, Bäumler AJ (2017) Oxygen as a driver of gut dysbiosis. Free Radic Biol Med 105:93–101
9. Emerson JE, Stabler RA, Wren BW, Fairweather NFY (2008) Microarray analysis of the transcriptional responses of *Clostridium difficile* to environmental and antibiotic stress. J Med Microbiol 57(6):757–764
10. Neumann-Schaal M, Metzendorf NG, Troitzsch D, Nuss AM, Hofmann JD, Beckstette M et al (2018) Tracking gene expression and oxidative damage of O_2-stressed *Clostridioides difficile* by a multi-omics approach. Anaerobe 53:94–107
11. Giordano N, Hastie JL, Carlson PE (2018) Transcriptomic profiling of *Clostridium difficile* grown under microaerophillic conditions. Pathog Dis 76(2)
12. Weiss A, Lopez CA, Beavers WN, Rodriguez J, Skaar EP (2021) *Clostridioides difficile* strain-dependent and strain-independent adaptations to a microaerobic environment. Microb Genomics 7(12)
13. Kint N, Morvan C, Martin-Verstraete I (2022) Oxygen response and tolerance mechanisms in *Clostridioides difficile*. Curr Opin Microbiol 65:175–182
14. Nicolas P, Mäder U, Dervyn E, Rochat T, Leduc A, Pigeonneau N et al (2012) Condition-dependent transcriptome reveals high-level regulatory architecture in *Bacillus subtilis*. Science 335(6072):1103–1106
15. Govind R, Dupuy B (2012) Secretion of *Clostridium difficile* toxins A and B requires the Holin-like protein TcdE. PLoS Pathog 8(6): e1002727
16. Yanni D, Kalziqi A, Thomas J, Ng SL, Vivek S, Ratcliff WC, et al. (2017) Life in the coffee-ring: how evaporation-driven density gradients dictate the outcome of inter-bacterial competition. arXiv

Chapter 10

Dual Fluorescent Labeling for Bacterial Aggregation Assays

Anna M. Tingler, Erin C. Chard, Kate Psenka, and Melinda A. Engevik

Abstract

Clostridioides difficile is an anaerobic, spore-forming pathogen that causes severe gastrointestinal disease. During infection, *C. difficile* interacts with other gut microbes within the intestinal mucus layer, which potentially influences colonization and disease progression. Here, we present a fluorescence-based protocol to examine bacterial adhesion and aggregation between *C. difficile* and other gut bacteria, such as *Fusobacterium nucleatum*. In this method, *C. difficile* is labeled with the green fluorescent dye CFDA-SE (carboxyfluorescein diacetate succinimidyl ester) and *F. nucleatum* or other bacterial partners are labeled with the red fluorescent dye CellTracker™ Orange CMRA (2,6-Bis-(4-chloro-phenyl)-isonicotinic acid 4-(1-carboxy-3-methyl-butoxy)-phenyl ester.) Following labeling, bacterial cultures are co-incubated in aggregation buffer, and aggregation is assessed by fluorescence microscopy. Alternatively, the labeled bacteria can be applied to cultured epithelial cells to evaluate both aggregation and host adhesion. This approach enables direct visualization and quantification of interspecies interactions, providing valuable insights into how pathobionts exploit interbacterial adhesion to persist and thrive within polymicrobial intestinal communities.

Key words *Clostridioides difficile*, Aggregation, Adhesion, Fluorescence

1 Introduction

Clostridioides difficile is a spore-forming, toxin-producing pathogen that causes antibiotic-associated colitis and is a major healthcare burden worldwide [1]. Although often studied in isolation, *C. difficile* exists in polymicrobial communities in the intestinal mucus layer, and recent evidence suggests that its ability to colonize and form biofilms may be enhanced by cooperative interactions with other bacteria [2–16]. *Fusobacterium nucleatum*, an anaerobic Gram-negative bacterium commonly associated with periodontal disease and colorectal cancer, has been shown to adhere to and

Authors Anna M. Tingler and Erin C. Chard are co-first author to this chapter.

Supplementary Information The online version contains supplementary material available at https://doi.org/10.1007/978-1-0716-5328-9_10.

Kurt H. Piepenbrink and Jennifer M. Auchtung (eds.), *Clostridioides difficile: Methods and Protocols*, Methods in Molecular Biology, vol. 3046, https://doi.org/10.1007/978-1-0716-5328-9_10,

aggregate with other microbial species through adhesins, promoting biofilm formation and colonization [17–20]. Our previous work demonstrated that *F. nucleatum* directly adheres to *C. difficile* in intestinal mucus via the RadD adhesin, enhancing dual-species biofilm formation [6].

To directly visualize and quantify these bacterial interactions, we developed a dual fluorescent labeling protocol using CFDA-SE to label *C. difficile* and CMRA to label *F. nucleatum* and other anaerobic bacteria. This method allows for tracking of bacterial co-localization, aggregation, and spatial architecture within a mucin-rich environment. Fluorescence microscopy provides both qualitative and quantitative data on bacterial interaction dynamics, and this approach is broadly adaptable for studying other bacterial pairs. The protocol presented here enables researchers to dissect the molecular and physical basis of interspecies interactions in the gut and provides a valuable platform for studying how mucus-associated communities shape microbial behavior and host outcomes.

2 Materials

For each of the preparations below, reagents were sourced from the suppliers listed in Table 1.

2.1 Bacterial Media

2.1.1 Brain–Heart Infusion Supplemented (BHIS) Liquid Medium

1. Weigh 18.5 g of Brain–Heart Infusion (BHI) Powder (FisherSci #B11059) and add it to a 1-L glass bottle.
2. Add 1 g of yeast extract (FisherSci #BP1422500) and 0.1 g of L-cysteine (FisherSci #AAA1043518) to the BHI powder to make Brain–Heart Infusion supplemented (BHIS) media.
3. Add 500 mL of distilled water to the powder and mix thoroughly.
4. Autoclave at 30 min of exposure time to sterilize the medium.
5. Directly after autoclaving, place the medium in the anaerobic chamber with the cap off for 1–2 days to allow the sterile media to become fully anaerobic for optimal growth of *C. difficile*.

2.1.2 Cycloserine Cefoxitin Fructose Agar (CCFA) Selective Plates

1. To make 500 mL of CCFA, add 20 g of peptone (FisherSci #DF0118-17-0), 2.5 g of sodium phosphate dibasic (FisherSci #S375500), 0.5 g of potassium phosphate monobasic (FisherSci #BP3621), 1 g of sodium chloride (FisherSci #BP3581), 0.1 g of magnesium sulfate heptahydrate (FisherSci #BP2131), 3 g of fructose (FisherSci #AAA1771830), 1.5 mL of a 0.3% solution of neutral red (FisherSci #AC415491000), and 7.5 g of technical agar (FisherSci #DF0812179) to a 1-L glass bottle.

Table 1
Reagents

	Reagent	Manufacturer	Cat#
Brain–Heart Infusion Supplemented (BHIS) Liquid Medium	Brain–Heart Infusion Powder	BD Bioscientifics	237200
	Yeast Extract	MP Biomedicals	103303
	Cysteine	ThermoScientific	A10435.30
Cycloserine Cefoxitin Fructose Agar (CCFA) Selective Plates	Peptone	FisherSci	DF0118-17-0
	Sodium Phosphate Dibasic	FisherSci	S375500
	Potassium phosphate monobasic	FisherSci	BP3621
	Sodium chloride	FisherSci	BP3581
	Magnesium sulfate Heptahydrate	FisherSci	BP2131
	Fructose	FisherSci	AAA1771830
	Neutral red	FisherSci	AC415491000
	Technical agar	FisherSci	DF0812179
	Cycloserine	FisherSci	C11895G
	Cefoxitin	FisherSci	AAJ6689103
Fusobacterium Selective Agar Plates	NA	FisherSci	NC1461788
Aggregation buffer	Sodium chloride	FisherSci	BP3581
	1 M Tris	VWR	E199
	Calcium chloride	Sigma Aldrich	C1016
	Magnesium chloride monohydrate	ThermoScientific	223210010
	Sodium hydroxide	FisherSci	S318-100
Fluorescent dyes	Carboxyfluorescein diacetate succinimidyl (CFDA-SE)	Life technologies	C1157
	CellTracker Orange CMRA	FisherSci	C34551
	Hoechst	Sigma Aldrich	B2261
Other reagents	Ethylenediamine Tetraacetate acid (EDTA)	Fisher Bioreagents	BP2482-100
	PBS	Quick Silver	EB1200
	FluoroBrite DMEM	ThermoFisher	A1896701
	Invitrogen ProLong™ RapidSet™ Antifade Mountant	FisherSci	P38930
	Bacteria	NA	NA
	Human colonic cells	NA	NA
	DMSO	FisherSci	BP231-100
Materials	Anaerobic chamber	AS-150 Anaerobic Chamber	150-2-029
	Sterile toothpicks	Amazon	600 count bamboo toothpicks
	Centrifuge	NA	NA
	15-mL conical centrifuge tubes	Stellar Scientific	T15-101-PK

(continued)

Table 1
(continued)

Reagent	Manufacturer	Cat#
Inverted phase contrast microscope	Mitotic	AE2000
Fluorescent microscope	Zeiss Axio Scan	NA
Frosted microscope slides	FisherSci	12-544-2
Coverslips 24x50mm	FisherSci	12-541-032
Two-sided disposable plastic cuvettes	VWR	97000-586
Spectrometer	Biorad SmartSpec Plus Spectophotometer	NA
Plate reader	Biotek Synergy H1	NA
96-well non-tissue culture treated plates	VWR	734-2781

2. Add 500 mL of distilled water to the powder and mix thoroughly.
3. Autoclave at 30 min of exposure time to sterilize the medium.
4. Allow the agar to cool to touch, but not solidify.
5. Add 2.5 mL of a 10% cycloserine solution (0.1 g of cycloserine in 10 mL water, FisherSci #C11895G) and 0.5 mL of a 1.56% cefoxitin solution (0.0156 g of cefoxitin in 10 mL of water, FisherSci #AAJ6689103) to the cooled medium.
   (a) Optional: This solution can be sterile filtered if desired.
6. Mix the solution thoroughly.
7. Pour ~25 mL of the solution into petri dishes in a biosafety cabinet or under a Bunsen burner.
8. Allow the plates to solidify and store them in the refrigerator until they are used.

2.1.3 *Fusobacterium* Selective Agar Plates

1. Anaerobe Systems Premade *Fusobacterium* Selective Agar plates (FisherSci #NC1461788).

2.2 Fluorescent Dyes and Buffers

1. 5(6)-CFDA-SE; CFSE (Life technologies #C1157).
   (a) Excitation/emission: 492 nm/ 517 nm.
   (b) A 100 mM stock solution is made by dissolving 25 milligrams of powder in 448.4 μL of DMSO. Store 10-μL aliquots at −20 °C to avoid frequent freeze-thawing.
   (c) The final concentration for labeling bacteria is 10 μM CFDA.
2. Invitrogen CellTracker Orange CMRA (FisherSci #C34551).

   (a) Excitation/emission: 548 nm/576 nm.
   (b) Each tube contains 50 μg of CellTracker Powder. To make a 10 mM stock solution, add 9.1 μL of DMSO directly to the tube.
   (c) The final concentration for labeling bacteria is 10 μM CMRA.
3. Sterile anaerobic PBS solution (QuickSilver Powdered Buffer Packs of PBS Buffer Powder Cat #EB1200, autoclaved after rehydration with 1 L water). Place the PBS in the anaerobic chamber for several days to fully deoxygenate the medium.
4. Aggregation buffer solution: 500 mL volume.
   (a) Add 4.38 g NaCl (final concentration: 150 mM) (FischerSci #BP3581).
   (b) Add 500 μL of 1 M Tris–HCl buffer solution, pH 8 (final concentration 1 mM) (VWR #E199).
   (c) Add 5.5 mg $CaCl_2$ (final concentration 0.1 mM) (Sigma-Aldrich #C1016).
   (d) Add 5 mg $MgCl_2{\cdot}H2O$ (final concentration 0.1 mM) (ThermoScientific #223210010).
   (e) Fill to 500 mL with water.
   (f) Adjust the pH to pH 7.5 using 1 M NaOH.
5. Hoechst No. 33342 (bisBenzimide H 33342 trihydrochloride) (Sigma Aldrich Cat #B2261):
   (a) Make stock solutions by adding 10 mg of Hoechst to 1 mL of water. This will generate a stock of 16.23 mM.
   (b) Aliquot the Hoechst into 10-μL aliquots and store at −20 °C.

2.3 Additional Reagents

1. Bacteria; example *C. difficile* R20291, *F. nucleatum* subspecies *polymorphum* ATCC 10953.
2. Anaerobic Chamber; example: AS-150 Anaerobic Chamber SN, 150-2-029.
3. Sterile toothpicks, autoclaved.
4. Centrifuge capable of spinning 15-mL culture tubes.
5. Sterile 15-mL Conical Centrifuge Tubes, Stellar Scientific #T15-101-PK.
6. Inverted phase contrast microscope capable of examining bacteria at least 40x; example, Motic AE2000 LED Microscope.
7. Fluorescent microscope capable of examining bacteria at least 40x; example, Zeiss Axio Scan.
8. Frosted microscope slides (FisherSci #12-544-2).

9. Coverslips 24 × 50 mm (FisherSci #12-541-032).
10. Invitrogen ProLong™ RapidSet™ Antifade Mountant (FisherSci #P38930).
11. Two-sided disposable plastic cuvettes (VWR #97000-586).
12. Spectrophotometer capable of measuring optical density at 600 nm; example: Biorad SmartSpec Plus Spectophotometer.
13. Plate reader capable of measuring fluorescence; example: Biotek Synergy H1 plate reader.
14. 96-well non-tissue culture treated plates (VWR #734-2781).
15. Human colonic cells (example: HT29-MTX cells, T84 cells, Caco-2 cells, etc.) grown to confluence on poly-L-lysine-coated coverslips.
16. FluoroBrite DMEM (ThermoFisher #A1896701).

3 Methods

3.1 Prepare the Bacteria

1. In the anaerobic chamber, aliquot 10 mL of BHIS (brain–heart infusion supplemented with yeast extract and cysteine) medium to three 15-mL sterile conical culture tubes.
   (a) See the diagram in Fig. 1.
2. Using a sterile toothpick, remove a single colony of *C. difficile* from a previously grown CCFA plate and inoculate one 15-mL conical tube containing 10 mL of BHIS.
   (a) Note: If you are doing multiple bacteria for aggregation in a single day, inoculate 50 mL of BHIS with *C. difficile*.
3. Using a pipette tip, remove a single colony of *F. nucleatum* from a previously grown *Fusobacterium* selective plate and inoculate one 15-mL conical tube containing 10 mL of BHIS. *See* **Note 1**.
4. Perform the motions of inoculating with a sterile toothpick without bacteria for the third BHIS tube and this will serve as a media control.
5. Place lids on all tubes and put the rack in the 37 °C portion of the anaerobic chamber.
6. Incubate the three BHIS tubes anaerobically overnight at 37 °C.
7. The following day, aliquot 150 μL of each conical tube into labeled 0.5-mL microfuge tubes and remove them from the anaerobic chamber.
8. In a biosafety cabinet, transfer 10 μL of each culture on to a petri dish or 96-well plate lid and examine the bacteria on a microscope at 40x to ensure the cultures are pure.

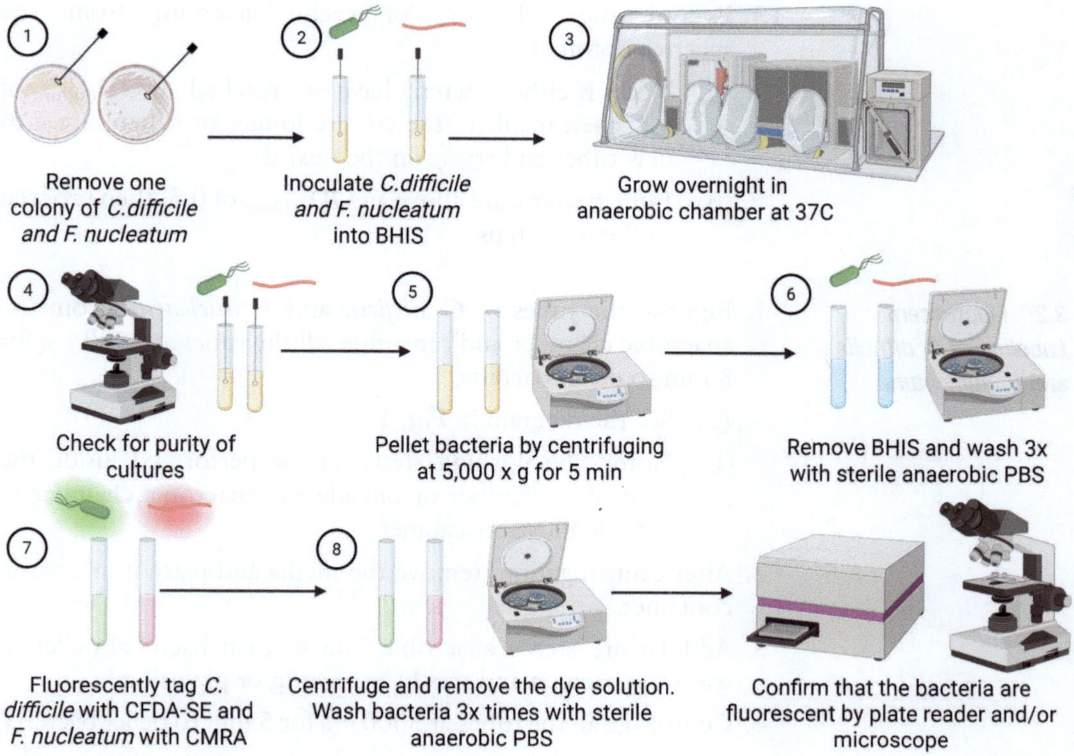

Fig. 1 Diagram of growing and fluorescently tagging the bacteria *C. difficile* and *F. nucleatum* with CFDA-SE and CMRA

(a) Note: *C. difficile* should appear as small rods that flip/spin and *F. nucleatum* should look like long thin spindles (Supplemental Fig. 1a, b).

(b) Ensure that there is no growth in the media control.

(c) If the cultures look okay, proceed to the next step.

9. In a biosafety cabinet, add 900 μL of PBS to 3x disposable plastic cuvettes.

10. Add 100 μL of each culture to an individual cuvette.

11. Add 100 μL of the media control to one cuvette; this will serve as your blank media control.

(a) Note: *C. difficile* and *F. nucleatum* generally grow to ODs higher than 2, which is the maximum level that most spectrophotometers can accurately measure. By making a 1:10 dilution of the bacterial culture in PBS, you can more accurately assess bacterial growth at an optical density of 600 nm (OD_{600nm}).

12. Use the media control sample as a blank to zero the spectrophotometer prior to measuring experimental samples.

13. Record the OD_{600nm} of each bacterium from the spectrophotometer.
    (a) Note: If either bacteria have not reached an OD_{600nm} of 0.5, then incubate the culture longer or subculture into new tubes and try again the next day.
    (b) If the bacteria are above an OD_{600nm} of 0.5, then proceed to the next steps.

3.2 Fluorescent Labeling of *C. difficile* and *F. nucleatum*

1. Remove the tubes of *C. difficile* and *F. nucleatum* from the anaerobic chamber and centrifuge all the tubes at 5000× *g* for 5 min to pellet bacteria.
    (a) See the diagram in Fig. 1.
    (b) Note: The washing steps can be performed inside the anaerobic chamber or outside the anaerobic chamber in a sterile biosafety cabinet.
2. After centrifugation, remove the media and place it in a waste container.
3. Add 10 mL sterile anaerobic PBS to each bacterial pellet to wash the bacteria and mix by vortexing or pipetting.
4. Centrifuge all the tubes at 5000× *g* for 5 min to pellet bacteria.
5. After centrifugation, remove the media and place it in a waste container. Repeat the washing step 2 more times.
6. Add 10 mL sterile anaerobic PBS to each bacterial pellet to wash the bacteria and mix by vortexing or pipetting.
    (a) Note: These washing steps are important to remove residual BHIS, which will interfere with the fluorescent tagging of the bacteria.
7. To the *C. difficile* tube, add 1 μL of the CFDA-SE stock (CFDA-SE stock: 100 mM) to make a final concentration of 10 μM CFDA-SE in PBS.
8. To the *F. nucleatum* tube, add 10 μL CMRA (CMRA stock: 10 mM) to make a final concentration of 10 μM CMRA in PBS.
9. Invert the tubes to mix the bacteria thoroughly with the dye.
10. Incubate the bacteria anaerobically at 37 °C for 1 h in the dark.
    (a) After the incubation, the liquid with CFDA-SE will appear slightly green and the liquid with CMRA will appear slightly pink (Supplemental Fig. 1c).
11. Following the incubation, centrifuge the tubes at 5000× *g* for 5 min.
12. After centrifugation, remove the staining solution and place it in a waste container.

13. Add 10 mL sterile anaerobic PBS to each bacterial pellet to wash the bacteria and mix by vortexing or pipetting.
14. Centrifuge all the tubes at $5000 \times g$ for 5 min to pellet bacteria.
15. Repeat the washing 2–3 additional times.
    (a) Note: The washing is very important because any remaining CFDA-SE or CMRA dye could stain co-incubating bacteria or cells.
16. After the final wash, resuspend the bacteria in 5 mL of sterile aggregation buffer if you are doing the aggregation assay or 5 mL of sterile PBS or cell culture medium, like FluoroBrite DMEM, if you are treating cells.
17. Confirm that the bacteria are fluorescent by examining them on a microscope or reading the fluorescence of the bacteria on a plate reader.
    (a) CFDA-SE: excitation/emission, 492 nm/ 517 nm (FITC channel).
    (b) CellTracker Orange CMRA: excitation/emission, 548 nm/ 576 nm (Cy3 channel).
18. Add 100 μL of the labeled bacteria to a cuvette containing 900 μL of PBS and using 100 μL of sterile aggregation buffer or PBS as a blank.
19. Record the OD_{600nm} of each bacteria.
20. Proceed to either the aggregation assay or the cell adhesion assay. *See* **Notes 3** and **4**.

3.3 Aggregation Assay

1. For the aggregation assay, calculate how much bacteria you need to reach an OD_{600nm} of 1 and OD_{600nm} of 2 in 1 mL for each culture.
   (a) Note: If you do not have a high enough OD_{600nm}, centrifuge the tube again, add 2.5 mL of aggregation buffer to concentrate the bacteria, and retake the OD_{600nm}.
2. In the anaerobic chamber, remove the appropriate number of 2-mL tubes and place them into a tube holder.
   (a) See the diagram in Fig. 2.
   (b) Note: Include three tubes for each bacterium alone and in combination to run the experiment in technical triplicate.
3. Add the appropriate amount of *C. difficile* to the tubes.
   (a) For the tubes where it is only *C. difficile*, add OD_{600nm} of 2 to each tube.
   (b) For tubes where it is *C. difficile* with another bacterium, like *F. nucleatum*, add OD_{600nm} of 1 of each bacterium to each tube to reach a total OD_{600nm} of 2.

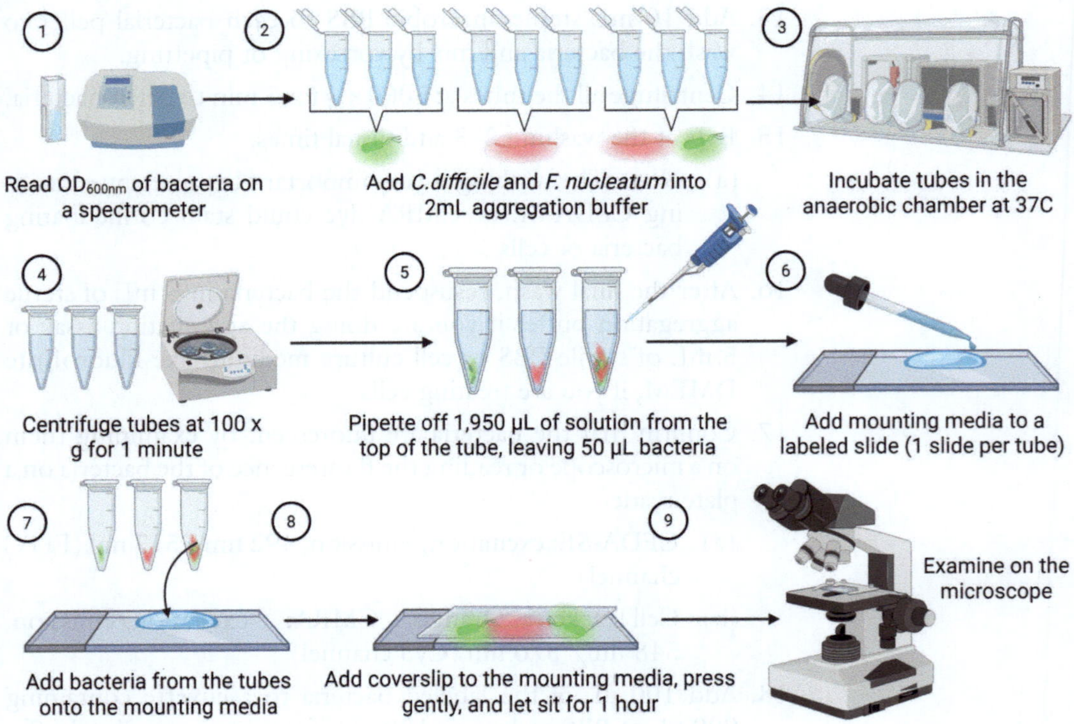

Fig. 2 Diagram of a bacterial aggregation assay with fluorescently tagged *C. difficile* and *F. nucleatum*

(c) For tubes where it's only *F. nucleatum*, add OD_{600nm} of 2 to each tube.

4. Add the appropriate amount of sterile aggregation buffer to each tube to reach a final volume of 2 mL.

   (a) Now all the tubes should now be at a total volume of 2 mL.

5. Incubate the tubes at 37 °C anaerobically for 1–6 h.

   (a) Note: The amount of time varies on the type of bacteria used. For *C. difficile* and *F. nucleatum*, 1 h is sufficient to examine aggregation.

6. After the incubation, remove the tube rack from the anaerobic chamber.

   (a) Note: Be careful not to disturb the tubes because you do not want to interrupt the aggregation.

7. Centrifuge the tubes at 100× *g* for 1 min to pellet co-aggregated cells while leaving non-aggregated bacteria in suspension.

8. Carefully pipette off 1.95 mL of the solution, leaving approximately 50 μL of bacteria behind.

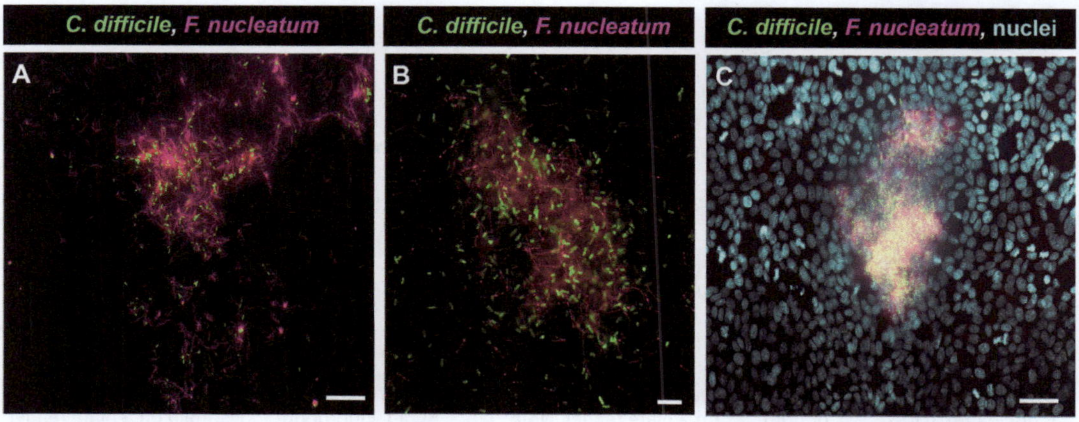

Fig. 3 Fluorescent microscopy of an aggregation assay with *C. difficile* tagged with CFDA-SE (green) and *F. nucleatum* tagged with CMRA (magenta) taken on a microscope at (**a**) 20x and (**b**) 100x. (**c**) Image of *C. difficile* and *F. nucleatum* aggregation on mucus-producing T84 cells where the T84 nuclei are stained with Hoechst (blue) at 20x

9. Label the appropriate number of glass slides (1 slide per tube) and add 3 drops of Invitrogen ProLong™ RapidSet™ Antifade Mountant.
10. In a biosafety cabinet, pipette up the remaining 50 μL of bacteria and add it to a labeled glass slide.
11. Add a coverslip on top and press gently to remove any air bubbles.
12. Allow the slide to set for at least 1 h and then image the bacteria on a fluorescent microscope.
    (a) See Fig. 3a for a 20x fluorescent image of *C. difficile* and *F. nucleatum* and Fig. 3b for a 100x fluorescent image of *C. difficile* and *F. nucleatum*.

3.4 Cell Adhesion Assay

1. Prior to this experiment, grow the cells of interest on poly-L-lysine-coated coverslips in a 24-well plate. These cells can be used when they are confluent.
    (a) See the diagram in Fig. 4. *See* **Note 2**
2. Transfer the 24-well plate containing confluent cells to a biosafety cabinet.
3. Aspirate the existing medium and wash the coverslips by adding 500 μL of sterile PBS to each well.
4. Aspirate the PBS to wash the cells.
5. Add 500 μL of FluoroBrite DMEM.
    (a) Note: FluoroBrite DMEM is optically clear and well suited for fluorescent applications.

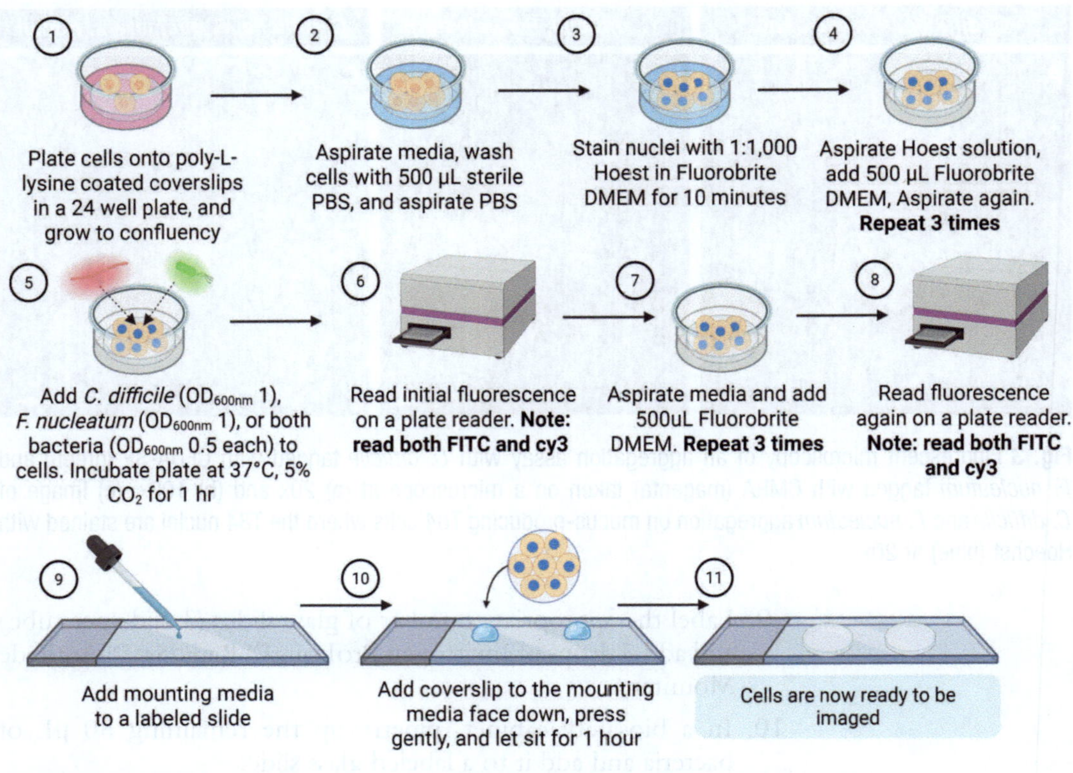

Fig. 4 Diagram of a cell adhesion assay with fluorescently tagged *C. difficile* and *F. nucleatum*

(b) Do not add fetal bovine serum to the adhesion assay as it can interfere with bacterial adhesion.

6. Stain the host cells with the nuclear dye Hoechst by making a 1:10 dilution of Hoechst stock (stock 10 mg/mL) in FluoroBrite DMEM to make a working solution of 1 mg/mL and add 5 µL of the diluted Hoechst to each well containing 500 µL of FluoroBrite DMEM (to make a final dilution of 1:1000 of Hoechst).
7. Incubate the cells at 37C, 5% CO_2 for 10 min.
8. After the incubation, aspirate the solution and add 500 uL of FluoroBrite DMEM to the well to wash.
9. Aspirate the remaining FluorBrite and add another 500 µL of FluoroBrite DMEM.
10. For the cell adhesion assay, calculate how much bacteria you need to reach an $OD6_{00nm}$ of 0.5 and $OD6_{00nm}$ of 1.
11. For the *C. difficile* alone wells, add $OD6_{00nm}$ of 1 of *C. difficile* to the wells.
    (a) Example: If you have *C. difficile* at OD = 3.52, use the following equation:

(b) Math: 1 (OD_{600nm} you want)/3.52 (OD_{600nm} of *C. difficile*) × 500 uL (volume per well) = 142. So remove 142 μL of FluoroBrite DMEM from the well and add 142 μL of *C. difficile* that is CFDA-SE tagged.

12. For the *F. nucleatum* alone wells, add $OD6_{00nm}$ of 1 of *F. nucleatum* to the wells.
    (a) Example: If you have *F. nucleatum* at OD = 4.8, use the following equation:
    (b) Math: 1 (OD_{600nm} you want)/4.8 (OD_{600nm} of *F. nucleatum*) x 500 uL (volume per well) = 104.2. So remove 104.2 μL of FluoroBrite DMEM from the well and add 104.2 μL of *F. nucleatum* that is CMRA tagged.
13. For the *C. difficile* and *F. nucleatum* combination wells, add $OD6_{00nm}$ of 0.5 of *C. difficile* and $OD6_{00nm}$ of 0.5 of *F. nucleatum* to the wells.
14. After the bacteria have been added, incubate the plate at 37 °C, 5% CO_2 for 1 h.
15. To get an assessment of how much bacteria were originally in the wells, read the entire 24-well plate on a plate reader.
    (a) CFDA-SE: excitation/emission, 492 nm/ 517 nm (FITC channel).
    (b) CellTracker Orange CMRA: excitation/emission, 548 nm/ 576 nm (Cy3 channel).
16. Then wash the wells by aspirating the medium and adding 500 μL of FluoroBrite DMEM to each well.
17. Aspirate the medium and add another 500 μL of FluoroBrite DMEM to each well.
18. Aspirate the medium and add another 500 μL of FluoroBrite DMEM to each well.
19. To get an assessment of how much bacteria are on the cells after washing, read the entire 24-well plate on a plate reader.
    (a) CFDA-SE: excitation/emission, 492 nm/ 517 nm (FITC channel).
    (b) CellTracker Orange CMRA: excitation/emission, 548 nm/ 576 nm (Cy3 channel).
20. Prepare the slides by labeling the slides and adding 2 drops of mounting medium to the upper and lower portion of the slide (Supplemental Fig. 1d).
21. Using forceps, gently remove the glass coverslips from the wells (Supplemental Fig. 1e, f).
    (a) Note: Keep the FluoroBrite DMEM media in the wells for this step; it will make it easier to remove the coverslips from the well.

22. Gently place the edge of the coverslip on a Kimwipe to remove any excess liquid.
13. Place the coverslip face down on a labeled glass coverslip that has a drop of mounting medium on it (Supplemental Fig. 1g).
    (a) Two coverslips can be placed on each slide.
23. Allow the slides to set for at least 1 hour and then image the slides on a fluorescent microscope.
    (a) CFDA-SE: FITC channel.
    (b) CellTracker Orange CMRA: Cy3 channel.

4 Notes

1. Aggregation assays can also be performed with other *Fusobacterium nucleatum* subspecies, such as *nucleatum*, *animalis*, or *vincentii* or other bacteria of interest.
2. Mucin glycans can be stained in the cell adhesion assay using fluorescently conjugated lectins: Jacalin (recognizes galactosyl (β-1,3) N-acetylgalactosamine, Vector Labs #FL-1151-5), SNA (recognizes α-2,6-sialic acid, Vector lab #CL-1303-1), or UEA-1 (recognizes fucose, Vector Labs #DL-1067-1).
3. Different components, such as mucus, dietary sources, etc., can be added to these experiments to identify the role of these compounds in bacterial aggregation and adhesion.
4. Bacteria can also be tagged using CellTracker™ Deep Red (#C34565, excitation 630 nm, emission 660 nm). This will allow researchers to incorporate three different bacteria, or add an additional stain, like a mucin glycan stain to the cell assay.

Disclosures All other authors have no competing interests to declare.

Author Contributions Concept and design (AMT, ECC, MAE), methodology (AMT, ECC, KP), drafting the manuscript (AMT, ECC), editing manuscript (KP, MAE), obtained funding (MAE).

Funding

This study was supported by grants from the National Institute of Health T32GM132055 (AMT), American Society for Investigative Pathobiology (ASIP) Summer Research Opportunity Program in Pathology (KP), and R35GM155451 (MAE).

References

1. Cheng JKJ, Unnikrishnan M (2023) Clostridioides difficile infection: traversing host-pathogen interactions in the gut. Microbiology (Reading) 169(2)
2. Semenyuk EG, Poroyko VA, Johnston PF et al (2015) Analysis of bacterial communities during Clostridium difficile infection in the mouse. Infect Immun 83(11):4383–4391
3. Engevik MA, Yacyshyn MB, Engevik KA et al (2015) Human Clostridium difficile infection: altered mucus production and composition. Am J Physiol Gastrointest Liver Physiol 308(6):G510–G524
4. Engevik MA, Engevik KA, Yacyshyn MB et al (2015) Human Clostridium difficile infection: inhibition of NHE3 and microbiota profile. Am J Physiol Gastrointest Liver Physiol 308(6):G497–G509
5. Furtado KL, Plott L, Markovetz M et al (2024) Clostridioides difficile-mucus interactions encompass shifts in gene expression, metabolism, and biofilm formation. *mSphere* 9(6): e0008124
6. Engevik MA, Danhof HA, Auchtung J et al (2021) Fusobacteriumnucleatum adheres to Clostridioides difficile via the RadD Adhesin to enhance biofilm formation in intestinal mucus. Gastroenterology 160(4):1301–1314. e1308
7. Sulaiman JE, Thompson J, Qian Y et al (2024) Elucidating human gut microbiota interactions that robustly inhibit diverse Clostridioides difficile strains across different nutrient landscapes. Nat Commun 15(1):7416
8. Lesniak NA, Schubert AM, Flynn KJ et al (2022) The gut bacterial community potentiates Clostridioides difficile infection severity. MBio 13(4):e01183–e01122
9. Smith AB, Jenior ML, Keenan O et al (2022) Enterococci enhance Clostridioides difficile pathogenesis. Nature 611(7937):780–786
10. Skraban J, Dzeroski S, Zenko B, Mongus D, Gangl S, Rupnik M (2013) Gut microbiota patterns associated with colonization of different Clostridium difficile ribotypes. PLoS One 8(2):e58005
11. Grinblat J, Weiss A, Grosman B, Dicker D, Beloosesky Y (2004) Diarrhea in elderly patients due to Clostridium difficile associated with Salmonella and Shigella infection. Arch Gerontol Geriatr 39(3):277–282
12. Schubert AM, Rogers MA, Ring C et al (2014) Microbiome data distinguish patients with Clostridium difficile infection and non-C. difficile-associated diarrhea from healthy controls. MBio 5(3):e01021–e01014
13. Liao F, Li W, Gu W et al (2018) A retrospective study of community-acquired Clostridium difficile infection in Southwest China. Sci Rep 8(1):3992
14. Fukugaiti MH, Ignacio A, Fernandes MR, Ribeiro Junior U, Nakano V, Avila-Campos MJ (2015) High occurrence of Fusobacterium nucleatum and Clostridium difficile in the intestinal microbiota of colorectal carcinoma patients. Braz J Microbiol 46(4):1135–1140
15. Sidner B, Lerma A, Biswas B et al (2023) Flagellin is essential for initial attachment to mucosal surfaces by Clostridioides difficile. Microbiol Spectrum 11(6):e02120–02123
16. Normington C, Moura IB, Bryant JA et al (2021) Biofilms harbour Clostridioides difficile, serving as a reservoir for recurrent infection. npj Biofilms and Microbiomes 7(1):16
17. Park J, Shokeen B, Haake SK, Lux R (2016) Characterization of Fusobacterium nucleatum ATCC 23726 adhesins involved in strain-specific attachment to Porphyromonas gingivalis. Int J Oral Sci 8(3):138–144
18. Kolenbrander PE, Andersen RN, Moore LV (1989) Coaggregation of Fusobacterium nucleatum, Selenomonas flueggei, Selenomonas infelix, Selenomonas noxia, and Selenomonas sputigena with strains from 11 genera of oral bacteria. Infect Immun 57(10): 3194–3203
19. Karched M, Bhardwaj RG, Asikainen SE (2015) Coaggregation and biofilm growth of Granulicatella spp. with Fusobacterium nucleatum and Aggregatibacter actinomycetemcomitans. BMC Microbiol 15:114
20. Nagayama M, Sato M, Yamaguchi R, Tokuda C, Takeuchi H (2001) Evaluation of co-aggregation among Streptococcus mitis, Fusobacterium nucleatum and Porphyromonas gingivalis. Lett Appl Microbiol 33(2):122–125

Chapter 11

In Vitro Measurement of *Clostridioides difficile* Biofilm Formation Induced by Gut and Microbiota-Derived Signals

Carolina Altaf Figueiredo and Bruno Dupuy

Abstract

Studying biofilm formation in liquid cultures is a rapid and straightforward method for identifying relevant experimental and bacterial factors before conducting tests in more complex gut models. Recent studies showed that several recurrent pathogenic bacteria form biofilms in response to host and microbiota-derived signals. *Clostridioides difficile* biofilms present as heterogeneous structures are highly dependent on strain, environmental signals, and surrounding microbes. Here, we describe protocols for studying induced *C. difficile* biofilm formation in the function of gut-derived factors.

Key words Biofilm, Microbiota, Bile acids

1 Introduction

C. difficile's biofilm formation is a complex process whose study has gained importance over the past decade. While major bacterial pathogens raise concerns about acquired antibiotic resistance, *C. difficile* also challenges antimicrobial treatments by its ability to produce biofilms that hamper antibiotic diffusion [1, 2]. As an anaerobe, *C. difficile* may benefit from biofilms by being isolated from the oxygen gradient of the gut, both at standard and inflammatory levels [3, 4]. In addition, biofilms keep vegetative cells senescent and protected from potential commensal microbial antagonists, whose role is highlighted by the success of fecal-matter transplantation in the treatment of severe *C. difficile* infection (CDI) [5]. Nevertheless, it has recently been suggested that the role of biofilm formation in *C. difficile* pathogenesis extends beyond primary infection and may be associated with CDI relapses [6]. Some clinical strains with high virulence and recurrence rates

Supplementary Information The online version contains supplementary material available at https://doi.org/10.1007/978-1-0716-5328-9_11.

Kurt H. Piepenbrink and Jennifer M. Auchtung (eds.), *Clostridioides difficile: Methods and Protocols*, Methods in Molecular Biology, vol. 3046, https://doi.org/10.1007/978-1-0716-5328-9_11,

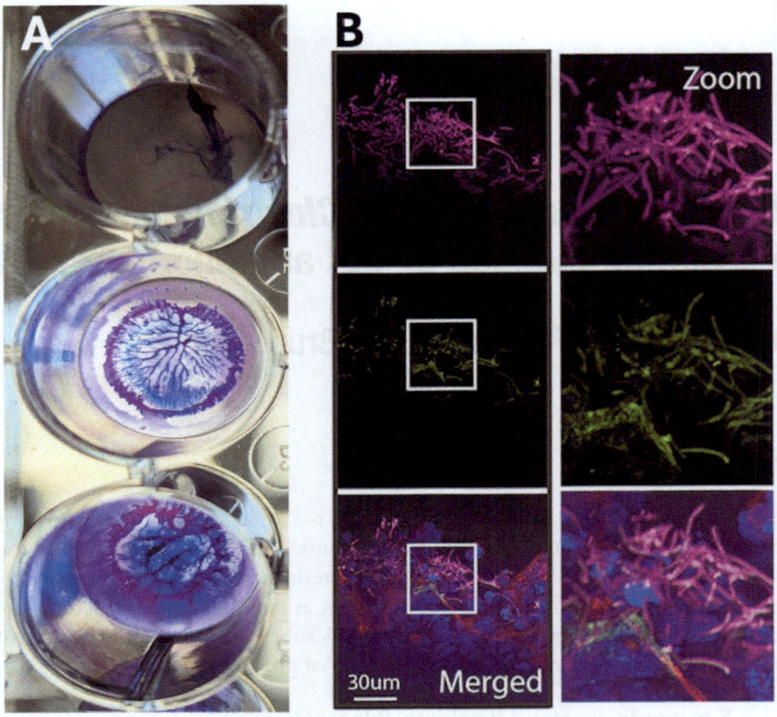

Fig. 1 In vitro induced *C. difficile* biofilms. (**a**) Biofilms induced by gut molecular signals in liquid culture, stained with 0.2% crystal violet. (**b**) *C. difficile* forms CDT-mediated microcolonies in an IoC model. Representative 3D reconstructed images of Caco-2 cells cocultured with HT29-MTX cells infected with *C. difficile* CNRS_CD129 (UK1 ToxAB−CDT+). DNA was labeled with DAPI (blue), mucin with anti-Muc5AC AF488 (green), actin with phalloidin rhodamine (red) and *C. difficile* with *anti-SlpA* AF647 (magenta). (Adapted from [10])

were identified as strong biofilm producers [7, 8]. *C. difficile* biofilm matrix mainly comprises extracellular DNA, exopolysaccharides, and proteins. While so far they seem qualitatively uniform, the abundance of each component can vary, which may result in changes to biofilm structure and live/dead cell content [8].

Most in vitro studies on biofilm formation are performed in liquid cultures on polystyrene plates [9], which allow for evaluating the biofilm biomass. The protocol of Dapa et al. (2013) was developed specifically to study *C. difficile* biofilm formation, resulting in the growth of confluent membranes (Fig. 1a). Overall, liquid culture assays are a prerequisite for optimizing conditions in studies of biofilm formation, particularly when we look at the largely unknown influence of gut metabolites on this process.

Advanced gut models, such as intestine-on-chip (IoC), Transwell, and vertical diffusion chambers (VDCs, [11]), are not specifically designed to study *C. difficile* biofilm formation, but serve as controlled alternatives before using animal models, which are highly complex [11]. As a result, the main observable signs of biofilm formation in these devices are increased bacterial adherence to epithelial cells and microcolony formation, that is, biofilm-like structures (Fig. 1b). Similarly, *C. difficile* forms aggregates in human intestinal organoids [12]. Lastly, in the mouse model, biofilms appear as 3-D islets/patches of bacterial aggregation at the cecal and colonic mucosa [13, 14]. Three-dimensional islets may be a result of the gut's irregular cellular topography, crypts, and fluid perfusion, which are not fully reflected in static and microfluidic models. Up to this day, there is no antibody designed for immunostaining *C. difficile*'s biofilm specifically, making the identification of biofilms in vivo difficult and rather qualitative. In this sense, factors identified from biofilms formed on liquid culture or advanced gut models can be integrated to visualize and interpret structures in vivo. However, the effectiveness of this approach remains constrained by model complexity and a lack of optimization. Thus, the use of specific biofilm formation assays is imperative to develop solutions for complex CDI models.

On the scope of simple in vitro biofilm formation, assays may be performed in liquid culture or agar. A study shows that both methods have biofilm matrix formation and differential gene expression compared to planktonic cell growth [9]. However, biofilms grown in liquid culture showed a more significant rearrangement of cellular metabolism in line with previous findings of biofilm transcriptomic studies [15]. Moreover, optimization of the latter is more applicable to the current complex models available (Transwell and organ-on-chip), due to the use of fluid confined to a surface [9]. Lastly, protocols on biofilm from liquid cultures were largely reviewed without substantial modification to date [8, 16, 17, 18].

In this chapter, we present protocols and their variations to study in vitro biofilm formation from liquid culture, in response to gut and microbiota-derivative signals optimized for both strong and fragile biofilms, as well as solutions to encountered issues.

2 Materials

2.1 Media for Biofilm Studies

The media of choice for biofilm formation vary according to the requirements of the conditions tested. One medium that is often used to compare both biofilm formation and induction between *C. difficile* strains is BHISG (brain–heart infusion [BHI] with 0.5% [w/v] yeast extract, 0.01 mg/mL cysteine, and 100 mM glucose) [19–22]. A protocol has also been optimized in Tryptone Yeast Extract medium (30 g/L tryptone, 20 g/L yeast extract) supplemented by 0.1% sodium thioglycolate (TYt), but only for non-induced biofilms [15]. Since biofilm formation by *C. difficile* remains a rather elusive process, the choice of a complex medium

Table 1
***C. difficile* semidefined biofilm medium component (BM)**

Component	Concentration (g/L)	Supplier
Oxoid casein hydrolysate	10	Oxoid
L-Tryptophan	0.5	
L-Cysteine	0.01	
L-Leucine	0.0033	
L-Isoleucine	0.0033	
L-Valine	0.0033	
Disodium phosphate(Na_2HPO_4)	5	
Sodium bicarbonate ($NaHCO_3$)	5	
Monopotassium phosphate (KH_2PO_4)	0.9	
Sodium chloride (NaCl)	0.9	
Ammonium sulphate ($[NH_4]2SO_4$)	0.04	
Calcium chloride dihydrate ($CaCl_2$-$2H_2O$)	0.026	
Magnesium chloride hexahydrate ($MgCl_2$-$6H_2O$)	0.02	
Manganese (II) chloride tetrahydrate ($MnCl_2$-$4H_2O$)	0.01	
Cobalt (II) chloride hexahydrate ($CoCl_2$-$6H_2O$)	0.001	
Ferrous (II) sulfate heptahydrate ($FeSO_4$-$7H_2O$)	0.004	
d-Biotin	0.001	
Calcium-d-panthothenate	0.001	
Pyridoxine	0.0001	

often depends on how closely the intended assay aligns with conditions reported to support biofilm development in the literature.

In induced biofilms, the choice of the medium depends on the molecular inducers used. For instance, DCA requires rich media such as BHISG to induce strong biofilm, while other inducers, such as pyruvate and succinate, form robust biofilm mainly in *C. difficile* semi-defined biofilm medium (BM) listed in Table 1 (*see* **Notes 1** and **2**), supplemented with 100 mM glucose (denoted BMG, *see* **Note 3**) [17, 20]. Both conditions promote high-biomass biofilms. However, BMG, as a minimal medium, makes it easier to identify metabolic processes and condition-specific metabolites of biofilm formation than BHISG medium. Nevertheless, it has been demonstrated that a rapidly metabolizable carbon source is necessary for robust biofilm induction [17, 20]. Even so, the extent to which supplements modulate biofilm formation varies according to the biofilm-forming capacity of each strain [8].

On a different note, when performing co-biofilms with *C. difficile* and another microorganism, the co-culture medium must be selected to allow growth of both species. Thus, when the BMG media is not applicable for a co-culture, supplementation with cysteine, glucose, or yeast extract (such as in BHISG) may support biofilm synthesis.

2.2 Supplies for Biofilm Formation and Fixation

For biofilm formation, 24-well plates flat bottom (polystyrene tissue culture-treated plates, Costar, USA) are preferred to that of 96-well plates (polystyrene tissue culture-treated plates, Costar, USA) as described below in methods. Materials for fixing biofilms vary based upon the strength of the biofilm formed as described in detail in Methods. For strong biofilms, fixation is achieved with a fixing solution of 25% v/v acetic acid and 75% v/v ethanol. For weak biofilms, fixation is achieved with Bouin's solution (Merck, HT10132; or 0.9% picric acid, 9% formaldehyde, and 5% acetic acid). Washing steps are performed with phosphate-buffered saline (PBS). For strong and weak biofilms, biofilm biomass is quantified with crystal violet solution (Merck, V5265) diluted in water to 0.2% or 1% w/v. Crystal violet is later dissolved in 100% ethanol (strong biofilms) or 25% v/v acetone and 75% v/v ethanol as described in Methods.

3 Methods

3.1 Experimental Setup

Whatever the medium used for biofilm assays, it must first be placed in the chamber to equilibrate to the temperature of 37 °C and the atmosphere ($H_2/CO_2/N_2$) of the incubator (*see* **Notes 4** and **5**). This avoids heat shock and toxicity from the oxygen usually present in the media recipients.

3.2 Overnight Cultures

The overnight cultures are performed in 5 mL of either BHI or TY media with antibiotics when appropriate. Media should be previously stabilized in the atmosphere and temperature of the anaerobic chamber and inoculated with a well-isolated bacterial colony. It is advisable to inoculate the media from freshly streaked colonies to ensure the viability of cell activity.

3.3 Biofilm Assays

Inoculation of plates for biofilm assays must be carried out carefully inside the anaerobic chamber to avoid contamination. The use of flat-bottom 24-well plates (polystyrene tissue culture-treated plates, Costar, USA; *see* **Note 6**) is preferred to that of 96-well plates (polystyrene tissue culture-treated plates, Costar, USA). The larger surface area of the 24-well plates may support thicker, mature biofilms, higher biomass (for downstream analysis such as nucleic acid extraction for omics study or imaging), and more medium (to avoid evaporation at longer incubation times). Nevertheless,

96-well plates are still relevant in the context of high-throughput screening, for which biofilm biomass is not an issue. In addition, the use of a flat rather than round bottom allows for uniform biofilm formation within the well, and therefore between wells, as well as compatibility with absorbance reading and microscopy devices.

It is necessary to distinguish constitutive biofilm formation, as observed for bacteria involved in most chronic infections in humans and biofilms induced and/or increased by the gut metabolic content. Dupuy and colleagues have recently identified metabolites present in the human gut whose concentrations oscillate according to gut health (i.e., dysbiosis), which induce *C. difficile* biofilm formation [17, 19, 20]. These biofilm inducers, such as deoxycholate (DCA) and succinate, differentially impact gene expression and result in biofilm with diverse matrix organizations [17]. Thus, *C. difficile* biofilms can be seen not only as a product of innate virulence but also as a reflection of the metabolic landscape of the gut. While genetic regulators of biofilm formation have been identified in major enteropathogens, deletion of their homologs in *C. difficile* does not completely abolish this phenotype [23–25].

1. All supplements tested for biofilm modulation (induction or inhibition) will be dissolved in pure water (acclimatized to the anaerobic chamber) and filtered through a 0.22-μm filter before being added to the culture media (*see* **Note** 7). Make sure to add an equivalent volume of filtered solvent to the negative control.
2. Inoculate 5 mL of the supplemented media (BHISG or BMG) with the overnight culture to have a final OD600 = 0.05.
3. Distribute 1 mL or 200 μL of inoculated medium over 4 wells of a 24-well and 96-well plate, respectively, to have four technical replicates per condition tested.
4. Incubate plates at 37 °C under anaerobiosis. While changes in the transcriptomic profile appear at 12–14 h of emerging biofilm, mature induced biofilms are mainly observed after 24–48 h of growth, followed by a progressive degradation of the biofilm biomass [8, 17, 20].
5. Shaking or major vibrations close to the plates over this step should be avoided, as they may disrupt the growing biofilms. Keeping a balanced humidity inside the anaerobic chamber (50–65%) prevents evaporation of the medium over long assays.

We generally observe high variability in biofilm biomass under the same conditions. That is because biofilms are highly susceptible to environmental variations, even when tested in the same settings, but days apart. This is especially true with *C. difficile* biofilms, which are often not easy to handle according to the strain used (see below). Thus, calculating the mean of four technical replicates and 4–5 biological replicates may be necessary to reduce standard error and assess significance, respectively.

3.4 Biofilm Fixation, Staining, and Biomass Quantification

C. difficile epidemic ribotype 027 strain is known for forming stronger biofilms than those of the *C. difficile* 630 strain, which are thin and frail, easily disrupted upon downstream processing [22]. However, the 630 strain has long been one of the main laboratory strains used to study *C. difficile*'s physiology, with constant updates on genome annotation and mutagenesis optimization [26]. Although all procedures remain the same regardless of the strain used, we have optimized the biofilm treatment for strong and fragile biofilms. We note that it is imperative to maintain the same treatment throughout the experiment to compare data.

Before treatment, avoid shaking the 24-well (or 96-well) plates when removing from the incubator. Inside an autoclavable plastic bag, invert the plate against a bundle of paper, in a rapid but gentle movement. This avoids spills and major disturbances of the biofilm layer.

1. Treatment for strong biofilms [17]:
   (a) Wash wells by pipetting gently 1 mL of PBS and remove it by inversion of the plate (here and on whenever removing any liquid from the plates).
   (b) Add 1 mL of a mix of 25% acetic acid and 75% ethanol (fixing solution) and incubate at room temperature for 20 min.
   (c) Remove the fixing solution and air dry for 1 h at room temperature.
   (d) Add 1 mL of a crystal violet solution (0.2%) to the wells and incubate for 10 min before removing it.
   (e) Wash the wells twice with PBS as described above, inverting the plate at the end (*see* **Note 8**).
   (f) Add 1 mL of ethanol (70%) to the wells.
   (g) Use a plate reader to read absorbance at the wavelength of 600 nm (optical density at 600 nm, OD_{600}), corresponding to biofilm biomass measurement (*see* **Note 9**).
2. Treatment for fragile biofilms (Altaf Figueiredo et al., in preparation; adapted from [27]):
   (a) Add 500 μL of the fixative Bouin's solution to the wells and incubate at room temperature for 15 min (*see* **Note 10**).
   (b) Remove the fixative and wash twice with PBS.

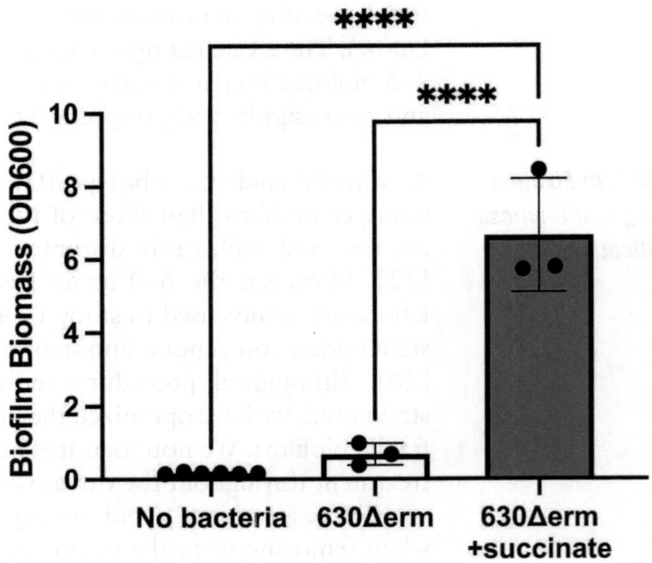

Fig. 2 Measurement of biofilm biomass in vitro following fixation with Bouin solution. The specificity of Bouin solution as a fixative was assessed after 48 h of incubation by applying the treatment for fragile biofilms (Altaf Figueiredo et al., in preparation) on BMG inoculated with sterile PBS (no bacteria), *C. difficile* 630Δ*erm* (*630Δerm*), and *C. difficile* 630Δ*erm* with succinate at 80 mM (*630Δerm + succinate*) after 48 h of incubation. Following fixation, biofilms were stained with 0.2% crystal violet (as described above), and biofilm biomass (OD_{600}) was measured. Statistics include one-way ANOVA, performed using GraphPad Prism version 11.0.0 for Windows, GraphPad Software, Boston, Massachusetts USA, www.graphpad.com

(c) After removing the PBS, add 500 μL of a crystal violet solution (0.2% or 1%, *see* **Note 11**) to the wells and incubate for 15 min before removing the staining solution.

(d) Wash the wells two or three times with PBS when using 0.2% or 1% crystal violet, respectively, inverting the plate at the end.

(e) Add 1 mL of a 25% acetone and 75% ethanol mix to the wells and incubate for 25 min at a plate-shaker at room temperature (*see* **Note 12**).

(f) Measure OD_{600}.

The purpose of Bouin's solution is to specifically stabilize biofilms, preventing them from being lost during downstream washing. As shown below, it does not result in residual staining of the culture media (Fig. 2, No bacteria; $OD_{600} = 0.174 \pm 0.009$ [SE]) and allows for a clear differentiation of baseline/low-biomass formation and high-biomass-induced biofilms (Fig. 2, *630Δerm* and *630Δerm + succinate*; $OD_{600} = 0.685 \pm 0.175$ [SE] and $OD_{600} = 0.647 \pm 0.897$ [SE], respectively).

4 Notes

1. Dissolve all ingredients from Table 1 in water and filter using a 0.22-μm filter.
2. Keep stocks of Oxoid casein hydrolysate, L-tryptophan, L-cysteine, L-leucine, L-isoleucine, L-valine, disodium phosphate, sodium bicarbonate, monopotassium phosphate, sodium chloride, d-biotin, calcium-d-pantothenate, and pyridoxine in aliquots at −20 °C to avoid waste when preparing small amounts of media (Supplementary Table 1). Once defrosted, do not freeze again. Remaining ingredients may be kept at 4 °C.
3. BM is a minimal medium whose stability is uncharacterized. Thus, to avoid degradation, always use BM freshly prepared, i.e., the day before.
4. BM medium may show precipitates/low opacity, which will disappear after acclimatization as the solutes dissolve. It should then be transparent with a white/yellowish glow. The presence of precipitates may cause variability in the results.
5. Media can be allowed to acclimatize for up to a minimum of 3 h before biofilm assays, but leaving it overnight allows checking potential media contamination.
6. Tissue culture-treated plates enhance cell attachment, while growth-promoting treatments may cause variability.
7. Optimal concentrations for using biofilm inducers have been defined for DCA (240 μM), pyruvate (100 mM), and succinate (80 mM) [17, 19, 20]. In all cases, the salt form of the molecules is used.
8. Both protocols may be interrupted and resumed the next day after washing the staining solution off the wells.
9. According to the plate-reader absorbance reading range, a 1:10 dilution of the 24-well plate contents is required at the end of steps 1.6/2.5. This allows for the quantification of high-biomass biofilms in individual wells that exceed the device's measuring capacities (>5).
10. Previously published protocols, such as the one for strong biofilms, use 1 mL of crystal violet when performing biomass quantification in 24-well plates. We chose to reduce the usage and discard of carcinogenic, mutagenic, and reprotoxic substances (CMRs), as *C. difficile* biofilms adhere to the bottom surface of polystyrene plates, which is completely covered with a 500 μL volume of either Bouin's solution or crystal violet [16].
11. 0.2% can be enough to observe biofilm formation, while 1% crystal violet may strengthen the staining.

12. High-biomass biofilms fixed with Bouin's solution often become opaque and yellow-colored membranes. This may interfere with the reading of absorbance. After the crystal violet is resuspended in the solubilizing agent, the contents of the biofilm assay cell-culture plate may be transferred to a new one for subsequent absorbance measurement.

References

1. Naghavi M, Stein DJ, Ikuta KS et al (2024) Global burden of bacterial antimicrobial resistance 1990–2021: a systematic analysis with forecasts to 2050. Lancet. https://doi.org/10.1016/s0140-6736(24)01867-1
2. James GA, Chesnel L, Boegli L et al (2018) Analysis of Clostridium difficile biofilms: imaging and antimicrobial treatment. J Antimicrob Chemother 73:102–108. https://doi.org/10.1093/JAC/DKX353
3. Lotoux A, Caulat L, Martins Alves C, Alves Feliciano C, Morvan C, Folgosa F, Martin-Verstraete I (2025) Defense arsenal of the strict anaerobe *Clostridioides difficile* against reactive oxygen species encountered during its infection cycle. MBio 16:e03753-24. https://doi.org/10.1128/mbio.03753-24
4. Dawson LF, Valiente E, Faulds-Pain A, Donahue EH, Wren BW (2012) Characterisation of *Clostridium difficile* biofilm formation, a role for *Spo0A*. PLoS One 7(12):e50527. https://doi.org/10.1371/journal.pone.0050527
5. Nooij S, Vendrik KEW, Zwittink RD, Ducarmon QR, Keller JJ, Kuijper EJ, Terveer EMT (2024) Long-term beneficial effect of faecal microbiota transplantation on colonisation of multidrug-resistant bacteria and resistome abundance in patients with recurrent *Clostridioides difficile* infection. Genome Med 16:37. https://doi.org/10.1186/s13073-024-01306-7
6. Meza-Torres J, Auria E, Dupuy B, Tremblay YDN (2021) Wolf in sheep's clothing: Clostridioides difficile biofilm as a reservoir for recurrent infections. Microorganisms 9(9):1922. https://doi.org/10.3390/microorganisms9091922
7. Martínez-Meléndez A, Morfin-Otero R, Villarreal-Treviño L, Baines SD, Camacho-Ortíz A, Garza-González E (2022) Analysis of biofilm production and expression of adhesion structures of circulating Clostridioides difficile strains from Mexico. Enferm Infecc Microbiol Clin (Engl Ed) 40(8):445–448. https://doi.org/10.1016/j.eimce.2021.01.008
8. dos Santos MMLG, Santos MGC, Costa CL, Martins CS, Leitão RFC, de Melo PD, Quesada-Gómez C, Castelo Branco D, Ferreira EO, Brito GAC (2022) Comparative biofilm-forming ability between *Clostridioides difficile* strains isolated in Latin America and the epidemic NAP1/027 strain. Front Cell Infect Microbiol 12:1033698. https://doi.org/10.3389/fcimb.2022.1033698
9. Brauer M, Lassek C, Hinze C, Hoyer J, Becher D, Jahn D, Sievers S, Riedel K (2021) What's a biofilm?—how the choice of the biofilm model impacts the protein inventory of Clostridioides difficile. Front Microbiol 12:682111. https://doi.org/10.3389/fmicb.2021.682111
10. Meza-Torres J, Tinevez JY, Crouzols A, Mary H, Kim M, Hunault L, Chamorro-Rodriguez S, Lejal E, Altamirano-Silva P, Groussard D, Gobaa S, Peltier J, Chassaing B, Dupuy B (2025) Clostridioides difficile binary toxin CDT induces biofilm-like persisting microcolonies. Gut Microbes 17(1):2444411. https://doi.org/10.1080/19490976.2024.2444411
11. Anonye BO, Hassall J, Patient J, Detamornrat U, Aladdad AM, Schüller S, Rose FR, Unnikrishnan M (2019) Probing Clostridium difficile infection in complex human gut cellular models. Front Microbiol 10:879. https://doi.org/10.3389/fmicb.2019.00879
12. Leslie JL, Huang S, Opp JS, Nagy MS, Kobayashi M, Young VB, Spence JR (2015) Persistence and toxin production by *Clostridium difficile* within human intestinal organoids result in disruption of epithelial paracellular barrier function. Infect Immun 83(1):138–145. https://doi.org/10.1128/IAI.02700-14
13. Soavelomandroso AP, Gaudin F, Hoys S, Nicolas V, Canny G, Bouillé A et al (2017) Biofilm structures in a mono-associated mouse model of Clostridium difficile infection. Front Microbiol 8:2084
14. Guo X, Wang J, Van Hensbergen VP, Liu J, Xu H, Hu X (2023) Interactions between host and intestinal crypt-resided biofilms are controlled by epithelial fucosylation. Cell Rep

42(7):112754. https://doi.org/10.1016/j.celrep.2023.112754

15. Poquet I, Saujet L, Canette A, Monot M, Mihajlovic J, Ghigo JM, Soutourina O, Dupuy B, Briandet R (2018) Clostridium difficile biofilm: remodeling metabolism and cell surface to build a sparse and heterogeneously aggregated architecture. Front Microbiol 9: 2084. https://doi.org/10.3389/fmicb.2018.02084
16. Đapa T, Leuzzi R, Ng YK, Baban ST, Adamo R, Kuehne SA, Scarselli M, Minton NP, Serruto D, Unnikrishnan M (2013) Multiple factors modulate biofilm formation by the anaerobic pathogen Clostridium difficile. J Bacteriol 195(3):545–555. https://doi.org/10.1128/JB.01980-12
17. Auria E, Tremblay YDN, Hamiot A, Martin-Verstraete I, Deschamps J, Briandet R, Dupuy B (2023) Extracellular succinate induces spatially organized biofilm formation in *Clostridioides difficile*. Biofilm 5:100125. https://doi.org/10.1016/j.bioflm.2023.100125
18. Abu Rahmoun L, Azrad M, Peretz A (2021) Antibiotic resistance and biofilm production capacity in *Clostridioides difficile*. Front Cell Infect Microbiol 11:683464. https://doi.org/10.3389/fcimb.2021.683464
19. Dubois T, Tremblay YDN, Hamiot A, Martin-Verstraete I, Deschamps J, Monot M et al (2019) A microbiota-generated bile salt induces biofilm formation in *Clostridioides difficile*. NPJ Biofilms Microbiomes 5:14. https://doi.org/10.1038/s41522-019-0087-4
20. Tremblay YDN, Durand BAR, Hamiot A, Martin-Verstraete I, Oberkampf M, Monot M, Dupuy B (2021) Metabolic adaption to extracellular pyruvate triggers biofilm formation in *Clostridioides difficile*. ISME J 15(12):3623–3635. https://doi.org/10.1038/s41396-021-01042-5
21. Doan T, Hoÿs S, Janoir C, Péchiné S (2022) Impact of subinhibitory concentrations of metronidazole on morphology, motility, biofilm formation and colonization of Clostridioides difficile. Antibiotics 11(5):624. https://doi.org/10.3390/antibiotics11050624
22. Pantaléon V, Monot M, Eckert C, Hoys S, Collignon A, Janoir C, Candela T (2018) Clostridium difficile forms variable biofilms on abiotic surface. Anaerobe 53:34–37. https://doi.org/10.1016/j.anaerobe.2018.05.006
23. Jackson DW, Suzuki K, Oakford L, Simecka JW, Hart ME, Romeo T (2002) Biofilm formation and dispersal under the influence of the global regulator CsrA of Escherichia coli. J Bacteriol 184(1):290–301. https://doi.org/10.1128/JB.184.1.290-301.2002
24. Yaron S, Römling U (2014) Biofilm formation by enteric pathogens and its role in plant colonization and persistence. Microb Biotechnol 7(6):496–516. https://doi.org/10.1111/1751-7915.12186
25. Püning C, Su Y, Lu X, Gölz G (2021) Molecular mechanisms of Campylobacter biofilm formation and quorum sensing. In: Backert S (ed) Fighting Campylobacter infections. Current topics in microbiology and immunology, vol 431. Springer, Cham, pp 293–319. https://doi.org/10.1007/978-3-030-65481-8_11
26. Smits WK (2017) SNP-ing out the differences: investigating differences between Clostridium difficile lab strains. Virulence 8(6):613–617. https://doi.org/10.1080/21505594.2016.1250998
27. Mihajlovic J, Bechon N, Ivanova C, Chain F, Almeida A, Langella P, Beloin C, Ghigo JM (2019) A putative type V pilus contributes to bacteroides thetaiotaomicron biofilm formation capacity. J Bacteriol 201(18). https://doi.org/10.1128/JB.00650-18

Chapter 12

Measuring *C. difficile* Adhesion to Mucosal Surfaces

Ben S. Sidner, Hugh C. McCullough, Baishakhi Biswas, Jennifer M. Auchtung, and Kurt H. Piepenbrink

Abstract

Colonization of the host gastrointestinal tract by *Clostridioides difficile* requires attachment to host tissues which is likely an integrative process potentially involving multiple modes of attachment, host cells, extracellular structures, and bacterial aggregation. Attachment to host mucosa represents a novel mechanism for understanding *C. difficile*'s persistence, either during infection or asymptomatic colonization. In this chapter, we describe methods for directly measuring the in vitro attachment of *C. difficile* to mucin hydrogels from a variety of sources.

Key words Mucin, Host-pathogen interactions, Glycobiology, Adhesion

1 Introduction

Clostridioides difficile [1] is an opportunistic, spore-forming pathogen capable of causing gastrointestinal illness (*C. difficile* infection or CDI). Recent studies have provided compelling evidence that microbe–mucus interactions, including attachment, are relevant to human CDI [2]. In a murine model of infection, *C. difficile* was found to associate with the loose outer mucus layer of the cecum and colon [3]. In that same study, Semenyuk and colleagues also found no evidence of colonization or attachment to epithelial cells, suggesting preferential colonization of the murine mucus layer by *C. difficile*. In CDI patients, expression of MUC2, the most abundant mucin glycoprotein in the human colon [4], decreased relative to healthy subjects [2]. Engevik and colleagues further demonstrated that *C. difficile* alone was capable of decreasing MUC2 expression in an infection model using human intestinal organoids. Altered mucin expression in CDI patients could therefore plausibly indicate compensatory downregulation. Additionally, using mucus extracted from CDI and healthy patient stools, *C. difficile* was found to preferentially adhere to CDI patient mucus. A later

Kurt H. Piepenbrink and Jennifer M. Auchtung (eds.), *Clostridioides difficile: Methods and Protocols*, Methods in Molecular Biology, vol. 3046, https://doi.org/10.1007/978-1-0716-5328-9_12,

study also reported preferential in vitro adherence to mucus from mucus-secreting human intestinal epithetical cell lines [5].

Here we describe methods for the creation of ex vivo mucosal hydrogels from purified mucin glycoproteins and techniques for the quantification of binding by *C. difficile* to those surfaces. These methods are general and can be used for both comparisons of mucosal adhesion between strains (naturally occurring or specific mutants) or evaluations of the conditions under which mucosal binding occurs (time, buffer conditions, etc.) [6].

2 Materials

2.1 Cell Lines

1. *C. difficile* strains (including ATCC BAA-1382).
2. LS174T cells (ATCC CL-188).
3. HT29-MTX-E12 Cells (Sigma #12040401; *see* **Note 1**).

2.2 Reagents

1. Brain–Heart Infusion Broth (Multiple sources, including Difco, Franklin Lakes, NJ and RPI).
2. Yeast Extract (multiple sources, including Fisher Bioreagents, Pittsburgh, PA, and Difco).
3. Phosphate-buffered saline (PBS, Fisher Bioreagents, Waltham, MA).
4. Thiamphenicol (Acros Organics, ThermoFisher Scientific, Waltham, MA).
5. Guanidinium chloride (multiple sources).
6. PMSF (phenylmethylsulfonyl fluoride, multiple sources).
7. CFDA-SE (5(6)-carboxyfluorescein diacetate succinimidyl ester; STEMCELL Technologies, Vancouver, Canada).
8. Porcine Gastric Mucin (Type III, Sigma-Aldrich, St. Louis, MO).
9. Trypsin–EDTA (ThermoFisher Scientific, Waltham, MA).
10. Ethanol (200 proof, multiple sources).
11. Acetone (Fisher, Waltham MA).
12. Sodium EDTA (multiple sources).
13. Sodium chloride (multiple sources).
14. Potassium phosphate monobasic and dibasic (multiple sources).
15. Iodoacetamide (multiple sources).
16. (3-Aminopropyl) triethoxysilane (Sigma-Aldrich, St. Louis, MO).
17. Sodium azide (multiple sources).

18. Glutamax (L-alanyl-L-glutamine dipeptide, ThermoFisher Scientific, Waltham, MA).
19. FBS (fetal bovine serum, multiple sources).
20. DAPT (STEMCELL Technologies, Vancouver, Canada).
21. DTT (dithiothreitol, Acros Organics, ThermoFisher Scientific, Waltham, MA).
22. Glutaraldehyde (Fisher, Waltham, MA).
23. cOmplete protease inhibitor (Roche, Basel, Switzerland).
24. Sodium borohydride (multiple sources).
25. Potassium chloride (multiple sources).
26. No. 1.5 glass coverslips (ThermoFisher Scientific, Waltham, MA). *See* **Note 2**.
27. Sterile 15-mL centrifuge tubes (Fisher, Waltham MA).
28. Sterile 50-mL centrifuge tubes (Fisher, Waltham MA).
29. Petri dishes (Fisher, Waltham MA).
30. Sterile loops (Fisher, Waltham MA).
31. Disposable tissue grinders (Fisher, Waltham MA).
32. 0.22-micron sterile filters (Multiple sources).
33. Coverslip stand (Millipore-Sigma Z688568).
34. 50-mM polypropylene petri dishes with lids (Fisher, Waltham MA).

2.3 Solutions

1. 60% EtOH (v/v) in deionized water.
2. 0.1 M NaCl pH 7.0 in autoclaved, deionized water.
3. 1 M HCl in autoclaved, deionized water.
4. Extraction buffer (6 M guanidium chloride, 5 mM NaEDTA, 0.01 M NaH_2PO_4, 100 mM PMSF, pH 6.5) in autoclaved, deionized water.
5. Reduction buffer (6 M guanidium chloride, 0.1 M Tris, 5 mM NaEDTA, 25 mM dithiothreitol, pH 8.0) in autoclaved, deionized water.
6. 4% solution of (3-aminopropyl) triethoxysilane in acetone.
7. 2.5% aqueous glutaraldehyde (v/v) in autoclaved, deionized water.
8. Mucin coupling buffer (5.6 g/L K2HPO4, 2.6 g/L KH2PO4, 8.7 g/L NaCl, 2.9 g/L NaEDTA, pH 7.2). Filter-sterilize through 0.22-micron filter.

2.4 Media

1. BHIY agar plates (37 g/L BHI medium, 15 g/L agar, 5 g/L yeast extract). Autoclave in a tray with water (at 120 °C for 30 min) (*see* **Note 3**).

2. BHIY broth (37 g/L BHI medium, 5 g/L yeast extract). Autoclave (at 120 °C for 30 min) (*see* **Note 3**).
3. Dulbecco's modified Eagle's medium (DMEM) with 2.5 mM Glutamax and 10% fetal bovine serum (FBS).

2.5 Equipment

1. Atomic force microscope (we used a Digital Instruments EnviroScope Atomic Force Microscope [ESCOPE] with a NanoScope V control station).
2. Centrifuge (we used a ThermoFisher Hereaus with TX-750 and JA-20 rotors).
3. Scanning electron microscope (we used a Hitachi FlexSEM 1000-II).
4. Stir plates (including one at room temperature and one in a refrigerated box or room).
5. Fluorescence microscope (we used an EVOS M5000).
6. Anaerobic chamber (Coy Lab Products).
7. Water baths or heating blocks.
8. Lyophilizer.
9. BSL-II laminar flow hood with UV light.

3 Methods

3.1 Culturing *C. difficile*

Overnight cultures of *C. difficile* can be grown from frozen glycerol stocks on BHIY agar plates at 37 °C in an anaerobic chamber (Coy Lab Products) with an atmosphere of 5% H_2, 5% CO_2, and 90% N_2. Thiamphenicol (15 mg/mL) or other antibiotics may be needed for plasmid maintenance.

1. Transfer glycerol stocks of the needed strains from the −80 freezer into a cold block and move them to the anaerobic chamber.
2. Streak out onto BHIY agar plates that have been pre-reduced in the chamber for ≥4 h.
3. After allowing 24–48 h for colony formation, examine the resulting plates for any potential contaminants. For experiments, individual colonies from overnight plates should be used to inoculate BHIY broth with appropriate antibiotics (*see* **Note 4**).
4. For adherence and microscopy experiments, 10 mL cultures in 15-mL centrifuge tubes generally provide sufficient sample.

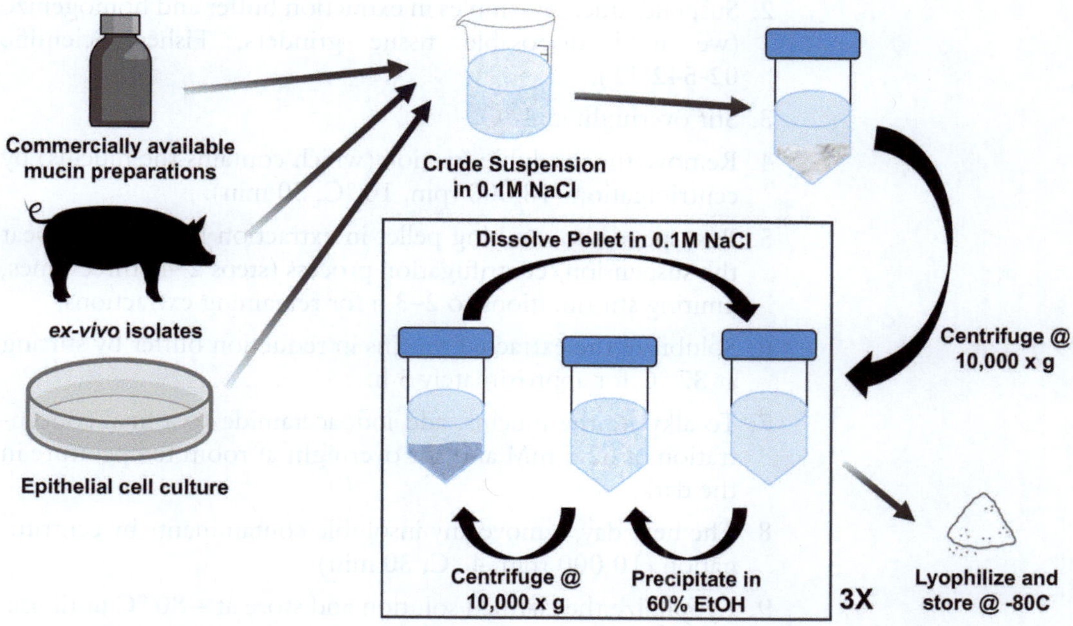

Fig. 1 Extraction and purification of mucin glycoproteins. The process for creating purified mucosal solutions from a variety of sources is schematically depicted

3.2 Partial Purification of Commercial Mucins

Commercial porcine gastric mucin (Type III, Sigma #M1778) should be partially purified prior to use [7, 8].

1. Suspend the crude mucin at approximately 20 mg/mL in 0.1 M NaCl.
2. Adjust the pH to 7.0 and stir overnight on a stir-plate at 4 °C.
3. The following day, pellet any insoluble residue by centrifugation (10,000× *g* for 10 min.).
4. Precipitate mucins contained in the supernatant by the addition of ice-cold ethanol to 60% (v/v) and then recover by centrifugation (10,000× *g* for 10 min.).
5. For purity, repeat the dissolution and precipitation steps twice more (See Fig. 1).
6. Resuspend the final pellet in 0.1 M NaCl pH 7.0.
7. Lyophilize the solution and store at −80 °C until use.

3.3 Extraction and Purification of Colonic Mucins

Colonic mucins can be purified from digestive tracts using a combination of mechanical and chemical purification [9, 10]. We have purified porcine colonic mucins from pig colons obtained from the University of Nebraska–Lincoln Department of Animal Science using the procedure outlined below.

1. Prepare the initial colonic mucosal tissue samples by gentle scraping of the mucosa with glass coverslips. Samples can be stored at −80 °C until further processing.

2. Suspend mucosal samples in extraction buffer and homogenize (we used disposable tissue grinders, Fisher Scientific 02-542-11).
3. Stir overnight at 4 °C.
4. Remove the insoluble fraction (which contains the mucins) by centrifugation (18,000 rpm, 10 °C, 30 min).
5. Resuspend the resulting pellet in extraction buffer and repeat the suspension/centrifugation process (steps 2–4) three times, limiting stir durations to 2–3 h for remaining extractions.
6. Solubilize the extracted mucins in reduction buffer by stirring at 37 °C for approximately 5 h.
7. To alkylate the mucins, add iodoacetamide to a final concentration of 62.5 mM and stir overnight at room temperature in the dark.
8. The next day, remove any insoluble contaminants by centrifugation (10,000 rpm, 4 °C, 30 min).
9. Lyophilize the clarified solution and store at −80 °C until use.

3.4 Expression, Extraction, and Purification of Mucins from Human Cell Culture

We grow immortalized cell lines in Dulbecco's modified Eagle's medium (DMEM) with 2.5 mM Glutamax and 10% fetal bovine serum (FBS) using a semi-wet interface culture method with mechanical stimulation as previously described [8].

1. Following passage, add DAPT to 10 mM to cells cultured in 150-mm tissue culture flasks containing 30 mL DMEM +10% FBS + Glutamax and incubate for 3–4 days in a 37 °C incubator with 5% CO_2 atmosphere until confluent (*see* **Note 5**).
2. After confluence is reached, replace media with 15 mL fresh DMEM and incubate with gentle rocking in a 37 °C incubator with 5% CO_2 atmosphere for three additional days.
3. Collect supernatants (containing mucins) and add protease inhibitors (1 mM PMSF, 2 mM iodacetamide), sodium azide (to 0.05%), and EDTA (to 10 mM). Shake for 1 h at 4 °C.
4. Precipitate mucins by the addition of 200 proof ice cold ethanol to 60% final concentration.
5. Centrifuge at 15,000× *g* for 30 min at 4 °C.
6. Wash the resulting pellets with 60% ethanol.
7. Resuspend in filtered, deionized water containing Roche cOmplete proteinase inhibitor cocktail, and store at −80 °C until lyophilization.
8. Lyophilize the solution and store at −80 °C until use.

3.5 Removal of O-Linked Glycans

Mucins are glycoproteins with both N-linked and O-linked glycosylation, but with O-linked glycans predominating. Some bacterial adhesion to mucins is independent of glycosylation [11], but we have previously shown *C. difficile* mucosal adhesion to be glycan-dependent [5, 6]. To remove O-linked glycans from porcine gastric mucin, we use β-elimination [12].

1. Resuspend lyophilized mucins in alkaline conditions (0.5 M $NaBH_4$, 50 mM KOH) at 50 °C overnight.
2. Cool sample on ice for 10 min.
3. Add ice-cold EtOH to 60% (v/v).
4. Centrifuge (10,000× *g* for 10 min) and resuspend in PBS (if lyophilizing and storing at −80 °C) or in mucin coupling buffer (if proceeding to preparation of artificial mucosal surfaces).

3.6 Preparation of Artificial Mucosal Surfaces

Any of the types of mucins described above can be covalently immobilized on glass coverslips using primary amine cross-linking [13, 14] as described below. Prior to cross-linking, we generally resuspend mucins in mucin coupling buffer at a final concentration of 1 mg/mL, although we have also tested other concentrations of mucins and used PBS as a buffer for mucin resuspension.

1. Wash cover slips with a mild, commercial dish detergent.
2. Wash thoroughly with deionized water.
3. Place coverslips in coverslip stand. Autoclave on dry cycle for 20 min (with 20 min of drying).
4. Place coverslips in 1 M HCl. Store at 60 °C overnight.
5. Wash coverslips in sterile deionized water.
6. Dry in oven for ≥1 h at 100 °C until fully dry (no liquid is visible). *See* **Note 6**.
7. Once dry, transfer coverslips to a 4% solution of (3-aminopropyl) triethoxysilane in acetone for 1 h, wash with acetone, and then heat treat at 100 °C for 1 h.
8. Incubate the coverslips in 2.5% aqueous glutaraldehyde (v/v) for 1 h. Rinse with water (*See* **Note 7**).
9. Transfer each coverslip to a polypropylene petri dish containing 3.5 mL of 1 mg/mL mucin in coupling buffer. Incubate overnight at 4 °C with gentle shaking.
10. Using sterile forceps in a biosafety cabinet, flip coverslips so that surface previously on air interface is now submerged in mucin-containing solution. Incubate with shaking at 4 °C for an additional 4 h.

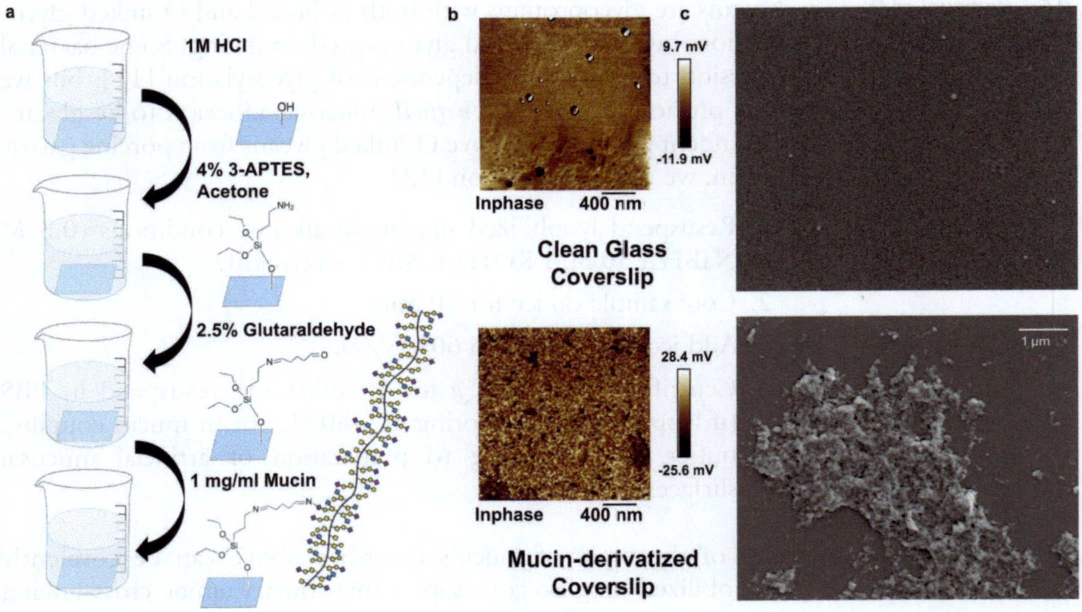

Fig. 2 Preparation of mucin-coated coverslips. The process for chemically modifying glass and attaching mucin glycoproteins is depicted on the left (**a**). On the right, (**b**) representative atomic force microscopy (AFM) and (**c**) scanning electron microscopy (SEM) data from derivatized and non-derivatized glass coverslips is shown

11. Air-dry and (for sterility) coverslips UV-treated in a biosafety cabinet, and store at 4 °C (We typically store in coverslip holders for ease of retrieval).
12. Successful deposition of mucosal hydrogels can be verified using atomic force microscopy (AFM) or scanning electron microscopy (SEM) (see Fig. 2).

3.7 Fluorescently Labeling and Visualizing Adhered C. difficile

Fluorescence microscopy can be used for direct visualization of *C. difficile* to coverslips, with or without derivatization [14].

1. Grow *C. difficile* broth cultures by selecting 5–10 colonies, inoculating in BHIY medium, and allowing to grow for 4–6 h to an OD600 of approximately 2.0.
2. Pellet through centrifugation (4500× *g* for 6 min). (*See* **Note 8**).
3. Gently resuspend cells in anaerobic PBS.
4. Repeat **steps 2** and **3** two more times to gently wash the cells.
5. Resuspend cells in anaerobic PBS containing 10 mM CFDA-SE and incubate in the anaerobic chamber at 37 °C for 1 h.
6. To remove any excess dye, centrifuge and resuspend in anaerobic PBS three times as described above.

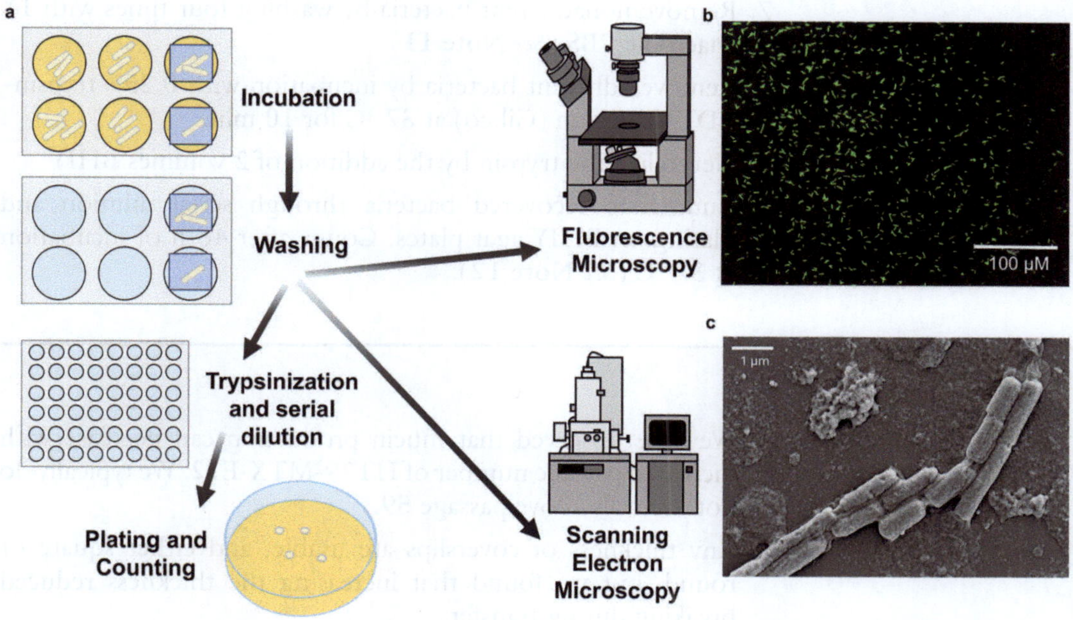

Fig. 3 Visualization and quantification *C. difficile* adhesion. The process for quantifying adhesion by CFU count is depicted on the left (**a**). On the right, (**b**) shows a confocal laser scanning microscopy (CLSM) image of fluorescently labeled *C. difficile* adhered to mucus, and similarly, (**c**) shows a scanning electron microscopy image of *C. difficile* adhered to mucus

7. Add 6 mL of labeled cells to a well of a six-well cell culture plates containing mucin hydrogel coated coverslips that have been pre-incubated in the anaerobic chamber.
8. Incubate anaerobically at 37 °C for 1 h.
9. Remove nonadherent bacteria by washing three times with 1x anaerobic PBS.
10. Remove coverslips from anaerobic chamber and fix in 2.5% glutaraldehyde for 10 min. Store at 4 °C until visualization.
11. Fixed, adherent, fluorescent bacteria can be visualized using standard methods (see Fig. 3).

3.8 Quantitative Mucosal Adhesion Assay

1. Grow *C. difficile* cultures in BHIY as described above, except that cells should be harvested at OD600 of approximate 0.5.
2. Centrifuge (5000× *g* for 5 min) (*See* **Note 9**).
3. Resuspend in anaerobic PBS to an OD600 of 0.5.
4. Add suspensions to six-well cell culture plates containing coverslips that have been preincubated in the anaerobic chamber (*See* **Note 10**).
5. Perform serial dilutions of inoculum in anaerobic PBS and quantify through plating on BHIY agar.
6. Incubate anaerobically at 37 °C for 1 h.

7. Remove nonadherent bacteria by washing four times with 1X anaerobic PBS (*See* **Note 11**).
8. Remove adherent bacteria by incubation with 0.25% trypsin–EDTA solution (Gibco) at 37 °C for 10 min.
9. Neutralize the trypsin by the addition of 2 volumes BHIY.
10. Enumerate recovered bacteria through serial dilution and plating on BHIY agar plates. Count after 48 h of incubation at 37 °C (*See* **Note 12**).

4 Notes

1. We have observed that mucin production can decrease with increased passage number of HT29-MTX-E12. We typically do not use cells above passage 59.
2. Any thickness of coverslips are usable, and either square or round, but we found that increasing the thickness reduced breaking during transfer.
3. BHIY is BHI medium supplemented with yeast extract, but without added cysteine. BHIY is used here to reduce levels of hydrogen sulfide production in the chamber, which can decrease the lifetime of O_2-scavanging palladium catalysts and other equipment in anaerobic chambers.
4. Groups of colonies (5–10) should be taken rather than individual colonies if there are concerns about genetic drift or genetic reversion mutants.
5. DAPT is a Notch pathway inhibitor which will induce differentiation of goblet cells.
6. Temperatures above 100 °C are not recommended because they can melt the coverslip holder.
7. To form terminal aldehydes on the surface.
8. A gentler centrifugation (1000 × *g* for 8 min) appears to preserve more flagella per cell. as does reduction of the acceleration and deceleration speed.
9. To reduce oxidation when the tubes are taken from the chamber for centrifugation, cell culture tubes can be wrapped in parafilm.
10. Coverslips should be transferred to 6-well plates with sterilized forceps in a biosafety cabinet to reduce risk of contamination. Cover slips are delicate, so care must be taken to handle gently with forceps positioned in the corner of the coverslip (to reduce removal of mucins/adherent cells) during all transfer steps.

11. For comparisons, the washing between groups must be equivalent. We have found that sequentially dipping the coverslips in sterile 50-mm petri dishes (or equivalent containers) of anaerobic PBS produces the most consistent results.
12. For commercial lyophilized mucus preparations, dialysis with low MW tubing (ex. ThermoFisher Snakeskin Dialysis Tubing, 7 kDA MWCO, #68700) can be used to remove residual small molecules (including glycan fragments).

References

1. Lawson PA et al (2016) Reclassification of Clostridium difficile as Clostridioides difficile (Hall and O'Toole 1935) Prevot 1938. Anaerobe 40:95–99
2. Engevik MA et al (2015) Human Clostridium difficile infection: altered mucus production and composition. Am J Physiol Gastrointest Liver Physiol 308(6):G510–G524
3. Semenyuk EG et al (2015) Analysis of bacterial communities during Clostridium difficile infection in the mouse. Infect Immun 83(11): 4383–4391
4. Johansson ME, Larsson JM, Hansson GC (2011) The two mucus layers of colon are organized by the MUC2 mucin, whereas the outer layer is a legislator of host-microbial interactions. Proc Natl Acad Sci USA 108 (Suppl 1):4659–4665
5. Engevik MA et al (2021) Mucin-degrading microbes release monosaccharides that chemoattract Clostridioides difficile and facilitate colonization of the human intestinal mucus layer. Acs Infect Dis 7(5):1126–1142
6. Sidner B et al (2023) Flagellin is essential for initial attachment to mucosal surfaces by Clostridioides difficile. Microbiol Spectrum 11: e0212023
7. Beighton D et al (1988) Increased degradative enzyme production by dental plaque bacteria in mucin-limited continuous culture. Microb Ecol Health Dis 1(2):85–94
8. Gargano AFG et al (2014) Mucin-based stationary phases as tool for the characterization of drug-mucus interaction. J Chromatogr A 1351:70–81
9. Asker N et al (1995) The human MUC2 mucin apoprotein appears to dimerize before O-glycosylation and shares epitopes with the 'insoluble' mucin of rat small intestine. Biochem J 308(Pt 3):873–880
10. Carlstedt I, Lindgren H, Sheehan JK (1983) The macromolecular structure of human cervical-mucus glycoproteins. Studies on fragments obtained after reduction of disulphide bridges and after subsequent trypsin digestion. Biochem J 213(2):427–435
11. Lillehoj EP et al (2024) Stenotrophomonas maltophilia provokes NEU1-mediated release of a flagellin-binding decoy receptor that protects against lethal infection. iScience 27(9): 110866
12. Wilkinson H, Saldova R (2020) Current methods for the characterization of O-Glycans. J Proteome Res 19(10):3890–3905
13. Landry RM et al (2006) Mucin-Pseudomonas aeruginosa interactions promote biofilm formation and antibiotic resistance. Mol Microbiol 59(1):142–151
14. Engevik MA et al (2021) Fusobacterium nucleatum adheres to Clostridioides difficile via the RadD adhesin to enhance biofilm formation in intestinal mucus. Gastroenterology 160(4): 1301–1314. e8

Chapter 13

An In Vitro Model for Studying Interactions Between Gastrointestinal Microbes and Planktonic and Sessile *Clostridioides difficile* Populations

Jason Pizzini, Hugh C. McCullough, Ben S. Sidner, Robert A. Britton, Kurt H. Piepenbrink, and Jennifer M. Auchtung

Abstract

Interactions between *Clostridioides difficile*, the gastrointestinal microbiota, and the host mucosal epithelium play important roles in governing the ability of *C. difficile* to colonize and cause disease. Several in vitro tools have been developed to investigate *C. difficile* physiology in the presence of microbial communities. In this chapter, we describe a model for studying *C. difficile*-mucin interactions in the presence of a complex microbiota using continuous flow bioreactors. This model can facilitate mechanistic studies of specific microbes and mucin structures important for *C. difficile* colonization, complementing findings from animal models. The approach presented here builds upon the preceding chapter's protocol for generating mucin hydrogels on glass slides and extends it to examine *C. difficile* adhesion to mucosal surfaces.

Key words Microbiota, Mucin, Colonization resistance

1 Introduction

The gastrointestinal microbiota is known to play an important role in governing the ability of *C. difficile* to colonize and cause disease. While many studies to characterize effects of the microbiota on *C. difficile* colonization have been performed in animal models, in vitro models have emerged as a complementary approach for characterization of *C. difficile* interactions with the microbiota [1–3]. The benefit of these models is that they can be operated under controlled laboratory conditions, which facilitates mechanistic studies of microbial interactions in the absence of variation introduced by host interactions. Fueled by observations that *C. difficile* interacts with the mucosal epithelium in vivo [4, 5] and can

Authors Jason Pizzini and Hugh C. McCullough have equally contributed to this chapter.

Kurt H. Piepenbrink and Jennifer M. Auchtung (eds.), *Clostridioides difficile: Methods and Protocols*, Methods in Molecular Biology, vol. 3046, https://doi.org/10.1007/978-1-0716-5328-9_13,

preferentially utilize mucin associated monosaccharides during infection [6] and that mucin-associated microbial communities may contribute to colonization resistance [7], there is growing interest in developing tools to study *C. difficile*–mucin interactions in the presence of the microbiota in vitro [8, 9]. The methods presented here represent a refinement of the approach originally described by Engevik et al. [8, 9], where the authors reported *C. difficile* binding was higher in the complex microbial communities formed on mucin hydrogel-coated glass coverslips compared to uncoated glass coverslips, demonstrating preferential colonization of mucin. These methods also build on those developed for the use of minibioreactor arrays (MBRAs) to culture complex microbial communities; a protocol for use of MBRAs has recently been published in the Journal of Visualized Experiments [10]. The methods presented in this chapter focus on binding assays performed on mucin hydrogels in the presence of a complex fecal microbiota. For detailed descriptions of mucin purification and hydrogel formation, please refer to the previous chapter on tools for studying *C. difficile* mucosal adhesion in this volume (Chapter 12).

2 Materials

2.1 Fecal Sample Processing for −80 °C Storage

1. Autoclavable laboratory spatulas (Eisco S41699, or similar).
2. Sterile 50-mL centrifuge tubes (Globe Scientific 6287 or similar).
3. Portable laboratory scale with sensitivity to 0.1 g (Ohaus 30456421 or similar).
4. Vortex mixer (Fisherbrand 02215414 or similar).
5. Phosphate-buffered saline (PBS, Fisher Scientific BP3991 or similar).
6. Sterile 1.7-mL microcentrifuge tubes (Corning 3207 or similar).
7. Sterile disposable laboratory spreaders, Pipetman, and barrier tips.
8. Low-temperature laboratory freezer (−80 °C).
9. Digital Dry Bath/Block Heater (Fisher 88-870-003 or similar).
10. Taurocholate Cefoxitin–Cycloserine Fructose Agar (TCCFA) plates [11].

2.2 Bioreactor Assembly and Operation (See Note 1)

1. Nine 100-mL reusable media storage bottles with 45-mm neck (Corning 1395-100 or similar); one 250-mL reusable media storage bottle with 45-mm neck (Corning 1395-250 or

similar); 500-mL reusable media storage bottle with 45-mm neck (Corning 1395-500 or similar); 2000 mL reusable media storage bottles with 45-mm neck (Corning 1395-2 L or similar).

2. A 15-position magnetic stir plate (Thermo Scientific Cimarec Poly 15 stirrer UX-04656-40 or similar).
3. Two peristaltic pumps with low flow rate capabilities, such as the Masterflex Ismatec IPC High-Accuracy Multichannel peristaltic pump with 12-channel drive (Avantor MFLX78006-24, or similar).
4. A heated anaerobic chamber with active dehumidification and sufficient dimensions to accommodate the bioreactor setup (90 × 80 × 80 cm [*l* × *d* × *h*]) is needed. An electrical power supply with sufficient capacity for two pumps and the stir plate. A pass-through port with butyl rubber stopper is strongly recommended, as this allows removal of waste from the chamber.
5. Nine sterile 0.22-μm syringe filters.
6. Epoxy (Loctite Instant Mix 5 min or similar).
7. Nine coverslip holders (Fig. 1): The coverslip holder can be 3D printed from polypropylene or similar biocompatible plastic from the stereolithography file available at doi:10.5281/zenodo.17435697.
   (i) Before its first use, the hollow center of the coverslip holder's stem requires an insert to increase density, preventing the holder from floating. A steel nail (#19 × ½ inch) can be trimmed and inserted to ensure the head is flush with the underside of the holder (see Fig. 1).

Fig. 1 Images of coverslip holder with six coverslips

(ii) The nail is then covered by epoxy to prevent leaching of iron.

(iii) The stem or a foot of the holder should be scored with a razor blade to facilitate consistent spatial orientation for coverslip sampling and replacement.

8. Mucin hydrogel-coated coverslips: prepared as described in Chapter 11, except that coverslips should be 10 mm square × 0.1 mm thick.
9. Autoclavable petri plates, such as Eisco #CH0372A.
10. Additional parts as listed in Table 1 below.
11. Cultivation medium plays a key role in determining which microbial communities assemble in bioreactors [12]. In previous studies monitoring *C. difficile* colonization, we have used BRM3 [13] medium. Each bioreactor requires approximately 1.5 L of BRM3 medium for a ten-day study. The recipe for 1 L of BRM3 is reproduced in Table 2 below.

2.3 Bioreactor Inoculation and Sampling

1. 25-mL sterile pipets.
2. PBS.
3. 2-mL sterile pipets.
4. 1.7-mL snap cap tubes.
5. 2-mL screw cap tubes.
6. 0.1-mm zirconia silicate beads (BioSpec Products 11079101Z or similar).
7. Nine 6-in or longer stainless-steel lab tweezers with broad tips.
8. Nine autoclavable P1000 pipet tip boxes with flat bottoms.
9. BeadBeater homogenizer (BioSpec Products 607 or similar).
10. UV Crosslinker (Stratalinker 1800 or similar).

3 Methods

3.1 Preparation of Fecal Samples for −80 °C Storage

1. Autoclave laboratory spatulas in aluminum foil or autoclavable pouches.
2. Place autoclaved spatulas, 50-mL centrifuge tubes, portable laboratory scale, vortex mixer, PBS, 1.7-mL microcentrifuge tubes, sterile spreaders, barrier tips, and TCCFA plates in anaerobic chamber to pre-reduce at least 24 h prior to receipt of fecal samples.
3. Upon receipt, transfer fecal samples to anaerobic chamber.
4. Use the sterile spatula to manually homogenize fecal material. Transfer 12-g aliquots to 50-mL centrifuge tubes and cap tightly. Remove from anaerobic chamber, label well, and store

Table 1
Tubing and fittings needed for assembly of nine reactors

Part number	Description	# pk	Image	Supplier
MFLX06424–67	C-Flex tubing; 25 ft/pk	2		Avantor
MFLX96460–26	2-Stop E-Lab Tubing, 0.89 mm; 12/PK	1		Avantor
MFLX96460–30	2-Stop E-Lab Tubing, 1.14 mm; 12/PK	1		Avantor
MFLX45505–82	Threaded male Luer, 25/PK	2		Avantor
MFLX45502–04	1/8″ female Luer; 25/PK	2		Avantor
MFLX45502–00	1/16″ female Luer; 25/PK	2		Avantor
MFLX45505–04	1/8″ male Luer; 25/PK	2	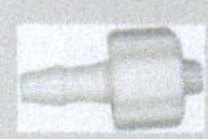	Avantor
MFLX45502–56	Female Luer tee; 1/pk	8		Avantor
MFLX45508–26	Female Luer elbow; 1/pk	2		Avantor
DM-MM-LUER-PP	Male to male Luer lock connector; 10/pk	1		Darwin fluidics
EW-21941-49	Bottle cap adapter; 5/pk	5		Fisher
EW-21943-04	Q-Series two-hole bottle cap (GL45); 1/pk	21		Cole- Parmer
EW-21942-76	PTFE tubing; 10 m/pk	1		Cole-Parmer
S99422	Pinchcock for tubing; 1/pk		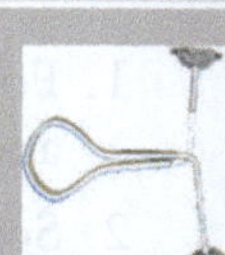	Fisher
UX-08546-00	Stir bar (10 × 3 mm)	9		Cole-Parmer

Table 2
Recipe for 1 L of BRM3 medium (*see* Note 2)

Reagent	Concentration	Heat stable
Bacto tryptone (Gibco 211705)	1 g	X
Bacto yeast extract (Gibco 212720)	2 g	X
Bacto Proteose Peptone No. 3 (Gibco 211693)	2 g	X
Bovine bile (Sigma B3883)	0.5 g	X
Sodium chloride	0.4 g	X
Porcine hemin solution (0.5% w/v in 1.4 M sodium hydroxide)	1 mL	X
Magnesium sulfate (1% w/v in distilled water)	1 mL	X
Calcium chloride (1% w/v in distilled water)	1 mL	X
Tween 80	2 mL	X
Arabinogalactan (from larch wood)	0.1 g	–
D-Cellobiose (sigma C7252)	0.15 g	–
Maltose	0.15 g	–
Inulin (from chicory)	0.2 g	–
D-Glucose	0.04 g	–
Sodium bicarbonate	2 g	–
Potassium phosphate dibasic (4% w/v in water)	1 mL	–
Potassium phosphate monobasic (4% w/v in water)	1 mL	–
Vitamin K1 (0.5% v/v in ethanol)	0.2 mL	–

at −80 °C. We have used aliquots of samples that have been stored at −80 °C for >3 years and observed similar results.

5. Remove 200-mg aliquot from remaining fecal material, transfer to 1.7-mL tube, add 800 μL PBS, cap tube, and vortex to resuspend.
6. Remove 200-mg aliquot from anaerobic chamber and heat treat at 65 °C for 30 min to kill vegetative cells.
7. Return 200-mg aliquot to anaerobic chamber and spread 100 μL on TCCFA plate. Incubate plate at 37 °C for 48 h to verify that fecal samples have undetectable levels of *C. difficile*.

3.2 Bioreactor Assembly (Nine Bioreactors; See Note 3)

1. For the nine two-hole bottle caps, remove the PTFE tubing and reserve it for further use.
2. Secure one bottle cap adapter into the top of each two-hole bottle cap.

3. In the second hole on the top of the two-hole bottle cap, secure one threaded male Luer fitting. Secure one 1/8″ female Luer into the male Luer fitting.
4. Cut nine segments each of PTFE tubing for interior waste and source lines of bioreactors. Source line tubing should be approximately 2.5 cm long and waste line tubing should be approximately 5 cm long.
5. Insert PTFE tubing inside the two-hole bottle caps. The waste line tubing should be inserted on the underside of the cap opposite the bottle cap adapter. The source line tubing should be inserted on the underside of the cap opposite the threaded male Luer fitting.
6. Fill nine 100-mL bottles with 80 mL distilled water (dH_2O) and coverslip holders.
7. Tighten two-hole bottle cap on each bottle to identify position at which waste line tubing sits at the air-water interface. Remove the bottle cap and use a razor blade to make a straight cut, trimming the waste line to the desired length that allows it to sit at the air–water interface.
8. Remove water and coverslip holder from each reactor. Insert micro stir bar.
9. Cut two 30-cm segments of C-flex tubing for each reactor. Insert one 1/8″ male Luer fitting into one end of each piece of tubing. Secure the other end of each piece of C-flex tubing to the fittings on the top of each two-hole bottle cap.
10. Gently heat ends of 2-stop pump tubing in a beaker of water containing dH_2O to increase flexibility. Insert 1/16″ female Luer fittings into the ends of each piece of tubing.
11. Connect 1/16″ female Luer on one end of 0.89-mm ID two-stop tubing to the 1/8″ male Luer inserted in the C-flex tubing that is connected to the source line port (shorter piece of PTFE tubing inside) of the two-hole bottle cap.
12. Connect 1/16″ female Luer on one end of 0.89-mm ID two-stop tubing to the 1/8″ male Luer inserted in the C-flex tubing that is connected to the source line port (shorter piece of PTFE tubing inside) of the two-hole bottle cap. Cover each of the 1/16″ female Luer fitting with aluminum foil for autoclaving. After autoclaving, these ends will be attached to the waste line tree assembled in **step 14**.
13. Cut a 15-cm piece of C-flex tubing for each reactor. Insert a 1/8″ male Luer fitting in one end and a 1/8″ female Luer fitting in the other end. Connect the 1/8″ male Luer fitting to the 1/16″ female Luer fitting at the other end of the 0.89-mm ID two-stop tubing. Cover the 1/8″ female Luer with a piece of aluminum foil for autoclaving.

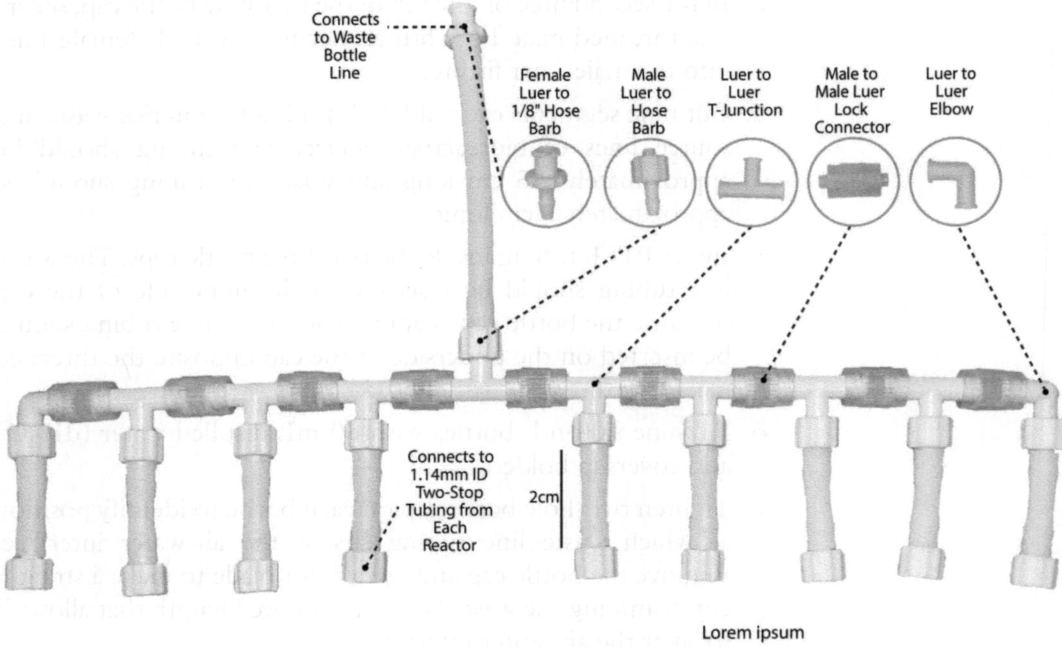

Fig. 2 Image of waste line tree. Image of the waste line tree set up for nine reactors, including tubing length and each individual part used for assembly

14. Assemble the waste line tree as shown in Fig. 2. Adjust the number of branching Luer tee junctions according to the number of reactors desired. Cover the exposed end with a piece of aluminum foil for autoclaving.
15. If anaerobic chamber has pass-through port, introduce a long piece of C-flex tubing through the port. Insert 1/8″ male Luer fitting into each end.
    (i) The length of this tubing will depend upon your specific laboratory setup. This tubing is to connect the waste tree to the waste receiver bottle cascade. The key principle is that the tubing should be of sufficient length to reach the waste receiver bottle cascade but should not be slack.
16. For setups with pass-through ports only, assemble the waste receiver bottle cascade. This cascade receives waste from the bioreactors in the chamber and is positioned in a large storage container that can contain any leaks (Fig. 3). A lid can be affixed to the top of the container, with access for tubing provided by drilling a small hole, to help contain smells associated with microbial waste.
    (i) Fit a 2-media storage bottle with a two-hole bottle cap fitted with two bottle cap adapters. Add an 8-cm piece of PTFE tubing to one of the bottle cap adapters. Waste will flow into the bottle via the bottle cap adapter while the

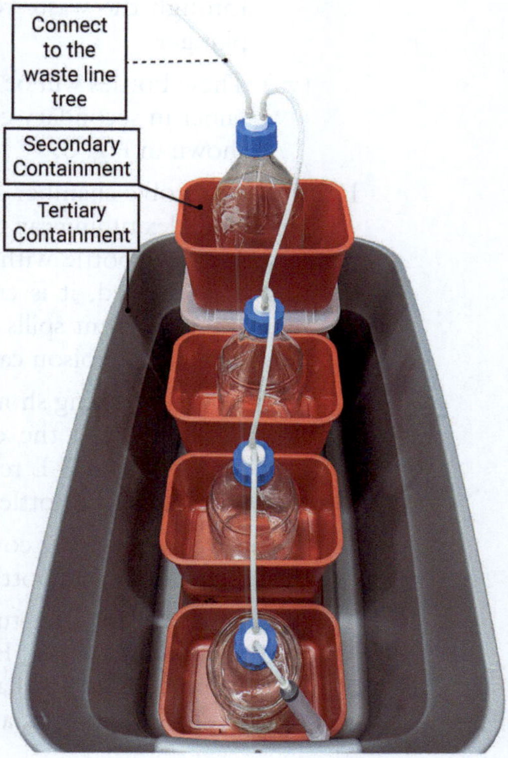

Fig. 3 Waste bottle cascade. An image of the cascading waste collection system. (This is adapted from Ref. [10])

PTFE tubing will transfer excess waste to the next waste bottle.

(ii) Cut two 12-cm sections of C-flex tubing and affix to the bottle cap adapters on the 2-L bottle. Insert a 1/8″ female Luer fitting into the end of the C-flex tubing that is on the bottle cap adapter that is not opposite PTFE tubing (see Fig. 3).

(iii) Fit a 2-L media storage bottle with a two-hole bottle cap fitted with an 8-cm piece of PTFE tubing opposite one of the bottle cap adapters.

(iv) Connect the piece of C-flex tubing from the first 2-L bottle to the bottle cap adapter that is not opposite the PTFE tubing. Cut a 12-cm section of C-flex tubing and affix to the remaining bottle cap adapter.

(v) Repeat step 16.iv two more times to assemble a series of bottles as shown in Fig. 3.

(vi) On the final bottle in the series, cut an 8-cm piece of C-flex tubing and affix one end to the bottle cap adapter. Insert a 1/8″ female Luer to the other end. A 20-gauge syringe can be optionally fitted to this port to pull waste through the waste receiver cascade by withdrawing the plunger.

(vii) These bottles will be positioned in the large storage container in secondary containment at decreasing heights as shown in Fig. 3.

17. If the anaerobic chamber does not have a pass-through port, a piece of C-flex tubing can be used to connect the waste tree to a waste receiver bottle within the anaerobic chamber. Note: If this setup is used, it is critically important to remove waste regularly to prevent spills and accumulation of hydrogen sulfide gas that can poison catalysts!
    (i) The C-flex tubing should contain a 1/8″ male Luer fitting in one end and the other end of the bottle should be connected to a 2-L receiver bottle fitted with a two-hole bottle cap and a bottle cap adapter.
18. Autoclave all reactors, coverslip holders (in pipet tip boxes), waste tree, and waste bottles for 30 min at 121 °C and 15 psi.
    (i) Loosely gather the tubing together with autoclave tape prior to autoclaving. Ensure that there are no sharp bends in tubing, as bends that are introduced during autoclaving will persist following autoclaving.

3.3 Assembly of Source Bottles and Preparation of Media (Nine Bioreactors)

1. Secure nine two-hole bottle caps to the tops of nine 2-L media storage bottles. The PTFE tubing that comes in the two-hole bottle cap should be of sufficient length to reach the bottom of the 2-L bottle.
2. Secure one bottle cap adapter to the top of the two-hole bottle cap opposite the PTFE tubing inside the bottle.
3. Secure one threaded male Luer into the second port of the two-hole bottle cap. Secure a 1/8″ female Luer fitting to the threaded male Luer.
4. For each source bottle, cut one 4-cm segment of C-flex tubing. Insert a 1/8″ male Luer into one end. Affix the other end of the tubing to the 1/8″ female Luer fitting on the two-hole bottle cap. Place a piece of aluminum foil over the male Luer fitting at the end of the 4-cm segment of C-flex tubing.
5. For each source bottle, cut one piece of 40-cm C-flex tubing. Insert a 1/8″ male luer fitting into one end. Affix the other end of the tubing to the bottle cap adapter on the top of the two-hole bottle cap.
6. For each source bottle, cut one 6-cm segment of C-flex tubing. Insert a 1/8″ female Luer fitting into one end and a 1/8″ male luer fitting into the other end. Connect the female Luer to the male Luer at the end of the 40-cm length of tubing. Cover the male Luer at the end of the 6-cm segment with aluminum foil.

7. Place a pinchcock on the 6-cm piece of tubing from step 6 approximately 1 cm from the male Luer fitting. Place the open loop of the pinchcock around the 4-cm segment of C-flex tubing that protrudes from the top of the two-hole bottle cap to gently bend the C-flex tubing.
   (i) This step prevents media from escaping during autoclaving.
   (ii) The pinchcock will damage the C-flex tubing during autoclaving, which is why this removable 6-cm piece of tubing is used during media preparation.
8. Prepare 14 L of BRM3 medium.
   (i) Add the heat-stable reagents in Table 1 to 13.3 L of dH_2O. Mix well.
   (ii) Remove two-hole bottle cap and transfer in 1425 of media to each source bottle. Remaining media volume can be preserved for *C. difficile* cultivation.
   (iii) Place two-hole bottle cap on top of each bottle and secure loosely (approximately 1/4 turn in threading that is reinforced with autoclave tape) to allow pressure release during autoclaving.
   (iv) Autoclave for 30 min at 121 °C with 15 psi.
   (v) While media with heat-stable reagents are being autoclaved, combine heat-labile reagents.
   (vi) Heat-labile reagents should be dissolved in 48 mL dH_2O per L of media, with pH adjusted to 7.2. To ensure sufficient volume for addition to medium with heat-labile reagents, we will typically prepare 5% excess. Heat-labile reagents are then sterilized by filtration through a 0.22-μ m filter.
   (vii) Allow autoclaved media to cool. Add 75 mL of filter-sterilized heat-labile reagents to each source bottle.

3.4 Preparation, Inoculation, and Sampling of Reactors

1. Prior to first use, bioreactor holders should be created, to ensure alignment of bioreactors with the stir plate. We have used one of the C-flex tubing boxes to create bioreactor holders that can be reused several times until replaced, as shown in Fig. 4. Three-dimensional-printed alternatives could also be created. Holes in cardboard were introduced with a razor blade that would position bottles over the centers of stirrers on the stir plate, with sufficient spacing to allow removal and sampling.
2. The two peristaltic pumps, stir plate, and bioreactor holder should be placed into the chamber prior to first use.

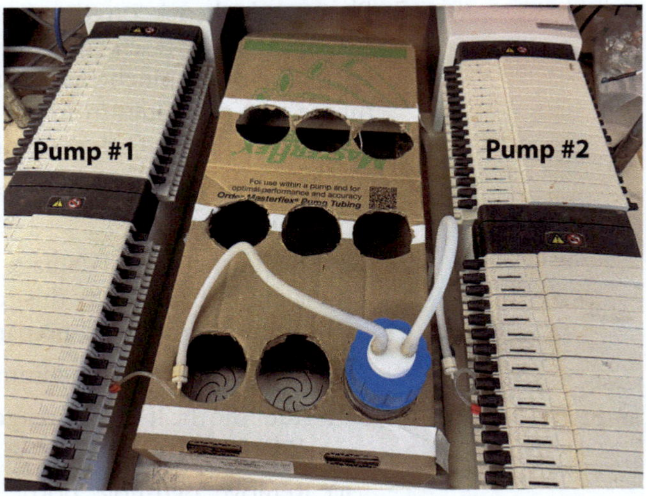

Fig. 4 Bioreactor setup. An image of the stir plate positioned between two pumps with a disposable bioreactor holder secured by laboratory tape. A single bioreactor is shown, but the setup can accommodate up to nine bioreactors

3. The peristaltic pumps are positioned on either side of the stir plate, with the stir plate centered between anaerobic chamber glove ports (Fig. 4).
   (i) We strongly encourage the use of wire shelves above the peristaltic pumps to allow media storage and more efficient usage of space.
4. Seven days prior to the start of the experiment, place source media bottles into anaerobic chamber to begin pre-reduction of media.
5. Two days prior to the start of the experiment, prepare the bioreactors and transfer into the anaerobic chamber.
6. Place autoclaved reusable petri dishes containing the prepared mucin-coated coverslips in a UV irradiator. Remove the lid and orient the coverslips with their surfaces facing the source. Irradiate at a setting of 120,000 joules/cm^2.
7. Close reusable petri plates and bring to biosafety cabinet. Use sterile technique to load mucin-coated coverslips into coverslip holders contained in autoclavable boxes using forceps.
8. Transfer boxes containing coverslip holders into the anaerobic chambers. Using forceps and sterile technique, introduce one coverslip holder with mucin-coated coverslips into each bioreactor.
9. Position bioreactors inside of the holders on the stir plate.
10. Secure the 0.89-mm two-stop tubing for each bioreactor into one channel of peristaltic pump #1.

11. Secure the 1.14-mm two-stop tubing for each bioreactor into one channel of peristaltic pump #2.
12. Carefully connect each female Luer fitting at the end of each waste line to the male Luer fitting at the of the C-flex tubing on the waste tree.
13. Connect the female Luer fitting at the end of the waste tree to the male Luer fitting that is connected to the waste bottle or tubing that connects to the waste receiver tree.
14. If using a waste receiver bottle cascade, connect the male Luer fitting at the end of the line that exits the anaerobic chamber to the female Luer fitting connected to the 2-L waste receiver bottle.
15. Connect each female Luer fitting at the end of the piece of C-flex tubing connected to the reactor by the 0.89-mm ID two-stop tubing to a male Luer fitting at the end of the source bottles.
    (i) Prior to making each connection, disconnect the segment of tubing containing the pinchcock.
    (ii) When removing aluminum foil, hold fitting near the hose barb to avoid contaminating connections that will be in contact with media.
    (iii) For gloved anaerobic chambers, sterile gloves can be worn over anaerobic chamber gloves.
16. Connect a 0.22-μm filter to the male Luer fitting that is at the end of the small segment of C-flex tubing on each source bottle.
17. Start the flow of peristaltic pumps #1 and #2 at a flow rate of 40 mL/h.
    (i) You should perform a test run of your setup with water prior to the first experiment to determine the appropriate pump settings to achieve a flow rate of 5 mL and 40 mL/h.
    (ii) A flow rate of 40 mL/hour will allow the reactors to fill to operating volume in 2 h.
    (iii) Run setup for 2.5 h to ensure that all reactors fill and there are no leaks in source or waste lines.
18. After 2.5 h, turn off pumps and allow to sit for 48 h to test for contamination and finish pre-reducing.
19. Place sterile PBS, pipets, 50-mL centrifuge tubes, autoclaved 250-mL bottle, 1.7-mL tubes, and pipet tips in chamber to pre-reduce. Make sure that closed containers are open to allow gas exchange.
20. On the day of the start of the experiment, thaw three tubes of fecal material in anaerobic chamber for approximately 20 min.

21. Add 36 mL PBS to each tube and vortex for 5 min. Seal well to prevent oxygen exposure and remove from chamber.
22. Centrifuge at 200× *g* for 5 min at room temperature to settle large particulates.
23. Return fecal slurries to anaerobic chamber.
24. Using aseptic technique, transfer approximately 36 mL of fecal slurry from each 50-mL tube to the 250-mL bottle. Close bottle and mix gently.
25. Transfer 16 mL to each reactor to inoculate. Avoid pipetting directly on the coverslip platform.
26. Start stirring of stir plates (approximately 250 rotations/min—a sufficient velocity to observe gentle mixing).
27. Leave inoculated reactors for 16 h for initial microbial outgrowth before starting flow of peristaltic pumps at 5 mL/h on pump #1 (source lines) and 10 mL/h on pump #2 (waste lines).
28. Remove all waste materials and disinfect. Re-autoclave forceps and tip boxes for sampling.
29. Sampling can be performed daily to assess microbial community development and to introduce fresh, mucin-coated coverslips.
    (i) For this process, pause the peristaltic pumps and stir plates, and then the two-hole bottle cap can be carefully removed from each bioreactor and secured to a sterile GL45 bottle. The lid from the sterile bottle should be placed on the bioreactor.
    (ii) Quickly remove the GL45 bottle lid and remove a 1-mL liquid sample with a sterile 2-mL pipet and transfer to sterile 1.7-mL tube. Replace lid.
    (iii) Remove the lid a second time and carefully remove the coverslip platform with sterile forceps and transfer to sterile empty pipet tip box with forceps. Replace the GL45 bottle lid on the bioreactor.
    - Take care when slowly raising the coverslip holder from the media to ensure that coverslips are square with the holder and not angled beyond the diameter of the bottle's mouth; otherwise, coverslips may break.

    (iv) Carefully remove two mucin-coated coverslips and transfer to two 50-mL centrifuge tubes. Introduce new mucin-coated coverslips.
    - Scoring the coverslip platform on the stem opposite one of the positions of the coverslip or on one of the feet will facilitate orientation for selection and replacement of coverslips with each sampling.

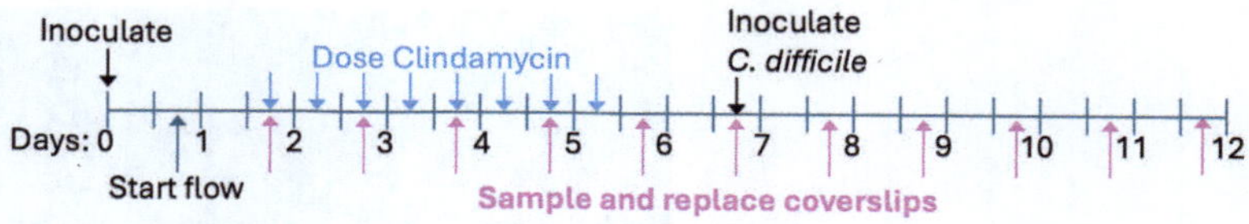

Fig. 5 Timeline for key steps in operation of bioreactors

(v) Return the mucin-coated coverslip platform to the bioreactor and then replace the two-hole bottle cap. Restart the peristaltic pumps and stir plates.

(vi) Remove planktonic and mucin-associated samples from the anaerobic chamber.

(vii) For planktonic cells, centrifuge in a microcentrifuge for 1 min at maximum speed to pellet cells and transfer supernatants to a new tube in the anaerobic chamber before storing both tubes at −80 °C.

(viii) For mucin-associated cells, store tubes at −80 °C until DNA extraction.

30. After 1 day with continuous flow, begin dosing reactors every 12 h with clindamycin at a final concentration of 500 μg/mL. Continue for 4 days (8 doses total). See Fig. 5.
31. On day 4, introduce TCCFA plate and a 15-mL centrifuge tube containing 10 mL BRM3 into anaerobic chamber to pre-reduce.
32. On day 5, streak *C. difficile* strain onto TCCFA agar plate and incubate anaerobically at 37 °C for 24 h.
33. On day 6, inoculate 5–10 *C. difficile* colonies into one BRM3 broth culture.
34. On day 6.6 (approximately 36 h after last clindamycin administration), introduce 800 μL of overnight culture into each reactor. Allow to mix, and then collect daily sample as described above.
35. Continue sampling for five additional days.
36. To analyze the presence of *C. difficile* in planktonic and sessile populations, measure with qPCR directly from cells that have been disrupted by bead beating with 0.1-mm BioSpec beads in 2-mL screw cap tubes as previously described [14].

    (i) To fit coverslips into screw cap tubes, you will need to break the coverslip into smaller pieces. Using sterile forceps, a clean break can be made by placing a coverslip against the side and base of a sterile tip box, effectively forming a right triangle with the coverslip serving as the hypotenuse. Then with the forceps partially open, press gently with prongs halfway across the coverslip as shown in Fig. 6.

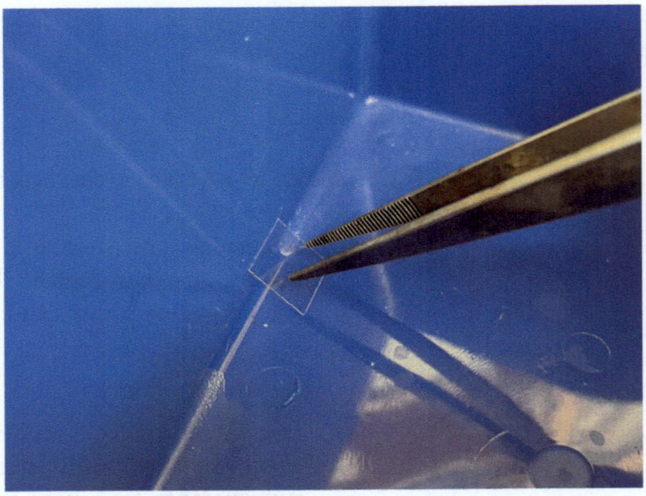

Fig. 6 Breaking glass coverslip with forceps prior to insertion into bead tubes

4 Notes

1. Methods are described for the operation of nine bioreactors, which is the preferred spacing based on space constraints and technical feasibility.
2. For any reagent without a specific supplier listed, we have had success with more than one trusted supplier of laboratory-grade reagents.
3. Assembly procedures for minibioreactor arrays (MBRAs), which share many similarities with the assembly procedures for these larger bioreactors, were recently described with video tutorials [10]. Readers are strongly encouraged to view this video protocol for clarification.

References

1. Wilson KH, Freter R (1986) Interaction of *Clostridium difficile* and *Escherichia coli* with microfloras in continuous-flow cultures and gnotobiotic mice. Infect Immun 54:354–358
2. Freeman J, O'Neill FJ, Wilcox MH (2003) Effects of cefotaxime and desacetylcefotaxime upon *Clostridium difficile* proliferation and toxin production in a triple-stage chemostat model of the human gut. J Antimicrob Chemother 52:96–102
3. Robinson CD, Auchtung JM, Collins J et al (2014) Epidemic *Clostridium difficile* strains demonstrate increased competitive fitness compared to nonepidemic isolates. Infect Immun 82:2815–2825
4. Semenyuk EG, Poroyko VA, Johnston PF et al (2015) Analysis of bacterial communities during *Clostridium difficile* infection in the mouse. Infect Immun 83:4383–4391
5. Engevik MA, Yacyshyn MB, Engevik KA et al (2015) Human *Clostridium difficile* infection: altered mucus production and composition. Am J Physiol Gastrointest Liver Physiol 308: G510–G524
6. Jenior ML, Leslie JL, Young VB et al (2017) Clostridium difficile colonizes alternative nutrient niches during infection across distinct

murine gut microbiomes. mSystems 2: e00063–17
7. Pereira FC, Wasmund K, Cobankovic I et al (2020) Rational design of a microbial consortium of mucosal sugar utilizers reduces *Clostridiodes difficile* colonization. Nat Commun 11: 5104
8. Engevik MA, Danhof HA, Auchtung J et al (2021) *Fusobacterium nucleatum* adheres to *Clostridioides difficile* via the RadD adhesin to enhance biofilm formation in intestinal mucus. Gastroenterology 160:1301–1314.e8
9. Engevik MA, Engevik AC, Engevik KA et al (2021) Mucin-degrading microbes release monosaccharides that chemoattract *Clostridioides difficile* and facilitate colonization of the human intestinal mucus layer. ACS Infect Dis 7:1126–1142
10. Pizzini JD, Midani FS, Britton RA (2025) Updated protocol for the assembly and use of the Minibioreactor Array (MBRA). J Vis Exp
11. Foster NF, Riley TV (2012) Improved recovery of Clostridium difficile spores with the incorporation of synthetic taurocholate in cycloserine-cefoxitin-fructose agar (CCFA). Pathology 44:354–356
12. McCullough HC, Song H-S, Auchtung JM (2025) Diversity in chemical subunits and linkages: a key molecular determinant of microbial richness, microbiota interactions, and substrate utilization. Microbiol Spectrum 0: e02618-24
13. Huang X, Johnson AE, Brehm JN et al (2025) *Clostridioides difficile* colonization is not mediated by bile salts and utilizes Stickland fermentation of proline in an in vitro model. 10:e01049-24
14. Auchtung JM, Robinson CD, Farrell K et al (2016) MiniBioReactor Arrays (MBRAs) as a tool for studying *C. difficile* physiology in the presence of a complex community. Methods Mol Biol 1476:235–258

Chapter 14

Conventional, Gnotobiotic, and Humanized Microbiota Mouse Models of *C. difficile* Infection

Daniel Erickson and James Collins

Abstract

Clostridioides difficile infection (CDI) remains a significant public health challenge, with murine models serving as a cornerstone for investigating host–pathogen interactions, microbiota dynamics, and therapeutic interventions. Although no single mouse model has become the standard, antibiotic-induced disruption of the gut microbiota remains the most common approach to facilitate *C. difficile* colonization. Here, we describe flexible and reproducible mouse models of CDI that can be adapted to diverse experimental goals, including studies of acute disease, long-term colonization, and microbiota-mediated resistance. This protocol includes detailed guidance on antibiotic pretreatment, *C. difficile* challenge, clinical monitoring, and microbial enumeration. We also outline procedures for microbial reconstitution in germ-free mice, enabling the study of defined or humanized microbiota. Practical considerations for minimizing environmental contamination and optimizing reproducibility are also discussed.

Key words Mouse model, Germ-free, Colonization

1 Introduction

Clostridioides difficile is a leading cause of healthcare-associated infections, with clinical manifestations ranging from mild diarrhea to life-threatening colitis. Although the role of *C. difficile* toxins in disease pathogenesis is well established, the complex interplay between pathogen, the host immune system, and the gut microbiota remains incompletely understood. Animal models have been instrumental in dissecting these interactions and in evaluating therapeutic strategies.

Among the available models, mice have emerged as the most widely used owing to their genetic tractability, availability of immunological tools, and compatibility with microbiome manipulation. However, murine models of *C. difficile* infection (CDI) are not standardized, and outcomes can vary depending on antibiotic pretreatment, microbial composition, and host genetics. Most models

Kurt H. Piepenbrink and Jennifer M. Auchtung (eds.), *Clostridioides difficile: Methods and Protocols*, Methods in Molecular Biology, vol. 3046, https://doi.org/10.1007/978-1-0716-5328-9_14,

rely on antibiotic-induced disruption of the gut microbiota to overcome colonization resistance, although germ-free and gnotobiotic systems offer alternative approaches for studying host–microbe–pathogen dynamics. Although antibiotic disruption is the most common method to induce susceptibility, alternative approaches, such as dextran sodium sulfate (DSS)-induced colitis or laxative treatment, can also increase susceptibility [1, 2].

Antibiotic regimens vary in intensity and composition, from a single administration of a single antibiotic to multiday multi-antibiotic regimens. Common single-antibiotic approaches involve antibiotics such as vancomycin, cefoperazone, and clindamycin, which are typically administered either orally via drinking water or gavage or as an intraperitoneal injection before infection [3–5]. More aggressive regimens include multiday oral administration of antibiotic cocktails followed by clindamycin injection [6, 7]. The diversity of available protocols reflects the need to tailor the experimental design to specific research questions. Different antibiotics influence the gut environment that *C. difficile* encounters and can result in distinct *C. difficile* carrier phenotypes [8, 9].

Germ-free mice offer a distinct model system in which colonization resistance is absent. In these animals, *C. difficile* readily colonizes the gut and results in a range of disease outcomes depending on the strain and inoculum used [10, 11]. Colonization and breeding of germ-free mice with human-derived microbiota (humanized-microbiota mice) can be used to study CDI with a microbiota more closely related to that of human patients and may enable the identification of beneficial microbes that are more directly relevant to the human population [12, 13]. Alternatively, colonization with defined microbial consortia, such as the altered Schaedler flora (ASF) or oligo-mouse-microbiota (OMM^{12} or $OMM^{19.1}$) may be used. These models often lack full colonization resistance and are susceptible to CDI without antibiotics [14–16].

This chapter provides a detailed modular protocol for establishing CDI in mice, including options for antibiotic pretreatment, microbial reconstitution, and infection monitoring. Practical considerations for reproducibility and biosafety are also discussed. This flexible framework is intended to support both mechanistic and translational research on *C. difficile* pathogenesis.

2 Materials

Note: This section describes all the materials listed in the following methods. Depending on the method chosen, not all materials may be necessary.

1. The selection of mice will depend on the study design and desired outcomes, as detailed in the *Methods* section.

2. When using a germ-free mouse to establish a humanized or gnotobiotic microbiota, a source of microbial inoculum, such as human feces or bacterial isolates, is essential.
3. Laboratory supplies including 50-mL Falcon tubes, 1.7-mL centrifuge tubes, 100-μm cell strainer, sterile phosphate-buffered saline (PBS), glycerol, 1% bovine serum albumin (BSA), 70% ethanol, CMDNT agar plates [17], vortex mixer, large-bore P1000 pipette tips, a sharps container, 96-well PCR plates and foil plate sealers, a thermocycler, 10% neutral buffered formalin (NBF), and clinical or United States Pharmacopeia (USP)-grade antibiotics as described in *Methods.*
4. Tools for working with mice including sterile 1-mL syringes, gavage needles, 23–29G × ½″ needle, Sharpie markers or ear tags, sterile and empty pipette tip boxes, a laboratory balance with precision to 0.01 grams, dissection-grade scissors and forceps, a polystyrene tray, pins, and tissue cassettes.

3 Methods

3.1 C. difficile *Infection*

Researchers should be trained in appropriate techniques such as intraperitoneal (IP) injection and oral gavage before starting.

3.1.1 Use of Defined or Humanized Microbiota (Optional)

If using conventional mice, proceed to *Setup*.

Many factors contribute to the successful colonization of germ-free mice; detailed protocols can be found elsewhere (e.g., [18]). We provide a brief protocol for microbial reconstitution of germ-free mice with human microbes (humanized microbiota), murine microbiota, or defined microbial communities.

1. Prepare microbial inoculum (e.g., human fecal slurry, murine cecal contents, or defined consortia) under anaerobic conditions. Sufficient inoculum should be prepared for all mice, with additional supplies available to account for unanticipated errors. An aliquot should be saved for sequencing to confirm the input inoculum.
   (a) If using a complex sample such as human fecal slurry or murine cecal contents, then dilute in 4 × volume of pre-reduced PBS in a sterile 50/mL Falcon tube.
   (b) Homogenize the sample by vortexing or pipetting up and down with a large-bore P1000.
   (c) Allow any large particles to settle for 10 min, and then carefully filter through a 100-μm cell strainer into a new sterile 50-mL Falcon tube. It is important that all particles that can block the gavage needle are removed.

(d) The sample can be used immediately, or a 1 × volume of pre-reduced 30% glycerol can be added and aliquoted into reduced freezer tubes and stored at −80 °C.

2. Orally gavage germ-free mice to reconstitute the microbiota (*see* **Note 1** and oral gavage details in "*C. difficile* challenge" section).
   (a) There is no consensus in the literature regarding the optimal dosing schedule. However, a good starting strategy to ensure robust colonization is to dose mice three times with 2–3 days between treatments.
   (b) Successful colonization can be confirmed via 16S rRNA gene sequencing or qPCR for smaller defined communities; however, care should be taken when using a targeted approach (qPCR) as possible contaminants will be missed.
3. House animals under strict gnotobiotic conditions to prevent contamination with other microbes [19].
4. Allow at least 7 days post-gavage for microbial stabilization before initiating antibiotic treatment or infection. Immune defects in germ-free mice have been well documented. If immune development is a concern, mice should be bred under gnotobiotic conditions, and the F1 and subsequent offspring of colonized breeders should be used (*see* **Note 2**).

3.1.2 Setup

1. If purchased from a vendor, the mice should be housed for at least a week upon receipt to allow for acclimation to the new environment (*see* **Note 3**).
2. Mark the tail of each mouse with a unique identifier using a sterile permanent marker (*see* **Note 4**). Reapply weekly or as needed. Alternative identification methods, such as ear tags, can be used.
3. Ensure that the mice are not colonized by *C. difficile* before experimentation (*see* **Note 5**). This can be achieved by selective plating or polymerase chain reaction (PCR) of unique markers such as toxin genes from mouse stool samples.
4. If microbiome analysis is planned, collect a baseline stool sample from each mouse into a sterile, pre-weighed microcentrifuge tube prior to any treatment and store at −70 °C until needed.

3.1.3 Antibiotic Treatment

Note: This step is not required for germ-free mice.

Where possible, clinical or United States Pharmacopeia (USP)-grade antibiotics should be used and may be required by your IACUC. Commonly used antibiotic treatments are outlined below.

(*A*) Single intraperitoneal (IP) injection of clindamycin [4]. (*B*) Five-antibiotic cocktail ad libitum in drinking water, followed by IP injection [6]. (*C*) Cefoperazone ad libitum in drinking water [5] (*see* **Notes 6** and 7).

A: Single IP injection of clindamycin

(a) Calculate the antibiotic concentration required to deliver 10 mg/kg clindamycin. If all mice have similar weights, the average weight can be used. If the weights differ significantly, for example, between males and females, set up two concentrations. Example:
   (i) Average mouse weight ≈ 25 g.
   (ii) Therefore, 0.25 mg clindamycin is required per mouse.
   (iii) If 200 μL of solution is administered to each mouse, clindamycin is diluted to 1.25 mg/mL in sterile saline.

(b) Fill 1-mL syringes (23–29G × ½″ needle) with 200 μL of the antibiotic solution (one per mouse, plus extra).

(c) Scruff the mouse securely and hold it in dorsal recumbency.

(d) Locate the peritoneal cavity in the lower right quadrant of the abdomen lateral to the midline (your left as you are looking at the upturned mouse). Avoid the left side to prevent penetration into the cecum.

(e) Insert the needle bevel-up into the peritoneal cavity at a shallow angle (30°–40°).

(f) Pull back on the plunger to ensure negative pressure prior to injection. If blood enters the syringe, the needle has not been inserted into the peritoneal cavity. Discard the syringe and repeat the procedure using a new needle.

(g) Administer the antibiotic slowly to minimize discomfort and dispose of the needle in a sharps container.

(h) Monitor the mice for any adverse reactions or signs of distress.

(i) Proceed to *C. difficile* challenge 24 h after intraperitoneal antibiotic injection.

B: Five-antibiotic cocktail plus clindamycin IP

(a) Prepare a fresh antibiotic mixture of kanamycin (0.4 mg/mL), gentamicin (0.035 mg/mL), colistin (850 U/mL), metronidazole (0.215 mg/mL), and vancomycin (0.045 mg/mL) in sterile water.

(b) Provide antibiotic mix to the mice ad libitum, replacing every 48 h to avoid antibiotic breakdown.

(c) Following 72 h of treatment, replace antibiotic water with fresh sterile water (*see* **Note 8**).

(d) Twenty-four hours after removing the antibiotic cocktail, administer clindamycin IP as described above.

(e) Proceed to the *C. difficile* challenge 24 h later.

C: Cefoperazone

(a) Prepare cefoperazone (0.5 mg/mL) in sterile water and administer to the mice ad libitum for 5 days (*see* **Note 8**).

(b) Replace antibiotic water every 48 h to prevent antibiotic breakdown.

(c) After 5 days of antibiotic treatment, replace cefoperazone with fresh sterile water.

(d) Proceed to the *C. difficile* challenge 24 h later.

3.1.4 *C. difficile* Challenge

All procedures should be performed in a biosafety cabinet within an ABSL2 facility.

1. Following antibiotic treatment, challenge the mice with spores of the chosen *C. difficile* strain via oral gavage (*see* **Note 9**).

   (a) Prior to beginning the experiment, prepare *C. difficile* spores suspended in 1% Bovine Serum Albumin (BSA) at the required dilution (*see* **Note 10**).

   (b) Attach a sterile oral gavage needle (typically 18-22G depending on the size of mice) to a sterile 1-mL syringe and fill the syringe with 200 μL of the spore solution (*see* **Notes 11** and **12**).

   (c) Restrain the mouse in a vertical position. Check the length of the gavage tube by measuring from the tip of the animal's head to the last rib. The gavage tube should not go beyond this point to avoid perforation of the stomach.

   (d) Maintain the animal in an upright (vertical) position. Advance the gavage needle along the roof of the palate, into the esophagus, and toward the stomach. If resistance is encountered, you may be attempting to enter the trachea, and the needle position should be altered. Once the needle has been inserted to a predetermined length, the inoculum is slowly injected into the stomach. Wait 2 s after injection and slowly remove the gavage needle from the mouse. If it appears that the material has been injected into the lungs (coughing or distress after removal), the animal should be euthanized.

3.2 Disease Assessment

1. Record baseline weight on the day of infection. An empty, sterile pipette tip box can be used to house individual mice while weighed.
2. Monitor mice daily for clinical manifestations of disease during the first 7 days and then every 2–3 days thereafter. Depending on the specific IACUC or model, mice may need to be monitored twice daily during acute infection or for a longer period. Disease progression can be monitored in two ways:
   (a) At a minimum, mice should be assessed for weight loss. Weights should be normalized to each mouse's baseline weight and represented as a percentage of the starting weight.
   (b) Use of a clinical disease scoring system. Several clinical scoring systems have been developed for this purpose. A scoring system optimized for *C. difficile* infection in mice was devised by Shelby et al. to assess stool consistency, behavioral changes, and weight loss [20]. When possible, a blinded third party should be present to perform clinical scoring.

3.3 C. difficile Enumeration

1. Collect fresh stool into pre-weighed microcentrifuge tubes. To obtain a stool sample, scruff the mouse and gently rub the stomach to induce defecation.
2. Reweigh the tubes and calculate the stool mass.
3. Transfer samples to the anaerobic chamber.
4. Prepare a 1:10 (w/v) dilution in pre-reduced PBS. Homogenize manually using a sterile pipette tip (*see* **Note 13**).
5. Allow larger stool sample particles to settle for 30 min in an anaerobic environment.
6. Prepare ten-fold serial dilutions in a 96-well PCR plate.
7. Spot 5 μL of each dilution in triplicate onto *C. difficile* selective pre-reduced CDMN agar [17] supplemented with 0.1% taurocholic acid (CDMNT agar) to enumerate total CFUs (*see* **Note 14**).
8. Seal the PCR plate, remove it from the anaerobic chamber, and heat-kill vegetative cells at 60 °C for 20 min in a thermocycler.
9. Return the PCR plate to the anaerobic chamber. Spot plate 5 μL of each serial dilution in triplicate onto pre-reduced CDMNT agar to enumerate *C. difficile* spores.
10. Incubate all plates anaerobically at 37 °C for 24–48 h.
11. Following incubation, remove spot plates from the anaerobic chamber and count colonies from the lowest dilution with discrete colonies.

(a) Use the following equation to enumerate *C. difficile* from each spot plate:

$$\text{CFU/g} = (\text{colonies}/[\text{dilution factor} \times \text{volume plated}])/(\text{stool weight/volume diluent})$$

(b) Use the following equation to enumerate vegetative *C. difficile*:

$$\text{Vegetative CFU/g} = \text{total CFU/g} - \text{spore CFU/g}$$

3.4 Necropsy

Necropsy procedures vary depending on the experimental goals. A basic procedure for harvesting colon and cecal tissue and its contents is outlined below. However, several in-depth necropsy protocols exist and should be considered before starting, for example, [21]. Necropsy should be performed once the mice reach euthanasia endpoints or the experiment reaches a predetermined endpoint. Tissues should be placed in the fixative immediately upon collection. Tissues most sensitive to autolysis (e.g., the GI tract) should be collected first, as degradation of the GI tract can begin within 10–15 min of euthanasia. If you are collecting samples from large groups of animals, you should euthanize them individually for tissue collection. Be sure to label all samples clearly with the animal ID, tissue type, and date.

1. Euthanize the mice according to AVMA guidelines and your IACUC-approved protocol. We use CO_2 asphyxiation, followed by cervical dislocation.
2. Following euthanasia, open the abdominal cavity of each mouse using dissection scissors soaked in 70% ethanol.
   (a) Lay the mouse on its back on top of a polystyrene surface or wax dissection tray. Secure the mouse by inserting a pin through each paw.
   (b) Spray liberally with 70% ethanol to disinfect and smooth down the fur.
   (c) Make a midline incision through the skin from the external genitalia to the tip of the chin.
   (d) Incise the skin along the hindlimbs. Pin back to expose the abdominal cavity.
3. Locate the colon near the rectum and gently pull up, snipping the colon as close to the anus as possible.
4. Gently pull the colon away from the body, allowing the intestine to unravel and detach from the mesentery.
5. Separate the cecum from the colon by cutting at the start of the proximal end of the colon and the distal end of the small intestine.

6. Cecal content can be collected by snipping the blind end of the cecal sac with sterile scissors and gently expelling the contents into a pre-weighed sterile 1.5-mL microcentrifuge tube. Reweigh the cecal contents and record weights. Flash freeze the cecal contents using liquid nitrogen. Store cecal contents on dry ice and transfer to −70 °C for long-term storage.
7. Flush with fixative (10% NBF) using a syringe and a 25-G needle.
8. Fix in 10% NBF for 24–48 h at 4 °C. Transfer to 70% ethanol for long-term storage at room temperature. This can be performed in labelled tubes or tissue cassettes.
9. Flush colons with ice-cold PBS (~8–10 mL) using a gavage needle and a 10-mL syringe inserted into the proximal end of the colon.
10. Cut into 1–2 cm segments or prepare a Swiss roll.
    (a) Gently flush with PBS.
    (b) Roll from the distal to the proximal end using forceps.
    (c) Place it in a cassette and fix it in 10% NBF. Keep colons in 10% NBF for 24–48 h and then transfer into 70% ethanol for long-term storage.
11. Place the carcasses into a biohazard bag and dispose of the carcasses according to the vivarium instructions.

4 Notes

1. For some microbes, colonization may be improved by combining oral gavage with colonic instillation [22].
2. If immune development is a concern, the F1 offspring of colonized breeders is the first generation that should be used. Future generations may also be used; however, 16s rRNA gene sequencing should be performed with each new generation of mice to ensure the microbial community is stable over time.
3. Our research group uses C57BL/6J mice obtained from Jackson Laboratories. Mice from different vendors or rooms within a vivarium may harbor distinct microbiota that influence CDI outcomes. Acquiring mice from the same commercial vendor and even the same room for independent experiments can improve the consistency of the results.
4. In our hands, Sharpie brand permanent markers fare the best after multiple rounds of autoclaving. However, autoclaving reduces the lifespan of the marker, regardless of the brand.
5. Mice should be screened for *C. difficile* prior to use. Some commercial mice may be latently colonized with toxigenic

C. difficile [23] which can affect experimental results and confer protection against severe CDI.

6. The choice of antibiotics should be considered when designing experiments. Different antibiotics can affect disease severity, and some may be more relevant in the human context. It is also important to note that the dose, duration, and antibiotic can affect the clearance of *C. difficile* following acute disease [9, 24].
7. Administration of antibiotics renders mice extremely susceptible to *C. difficile* infection. Given the propensity of *C. difficile* spores to persist in the environment, contamination and unintended disease within ~72 h following antibiotic treatment is extremely easy. We have found that the following precautions can reduce the likelihood of mice becoming contaminated with environmental *C. difficile* spores:
   (a) The workspace and closed cages should be sprayed and wiped clean using a bleach-based cleaner before and after handling the mice.
   (b) Nitrile gloves should be removed, and a fresh pair donned, between working with each individual cage of mice, or any time a surface outside the sterilized biosafety cabinet is touched.
   (c) Any material that comes into contact with the mice should be sterilized prior to use. This includes empty containers to weigh the mice and markers. Boxes are sterilized in an autoclave between each use, and the same box is used for its respective group for the entirety of the experiment.
   (d) If antibiotic-treated uninfected controls are part of the experimental design, all procedures involving these mice should be completed, and cages should be stored prior to opening any cages containing *C. difficile-infected* mice.
8. Antibiotic duration may require optimization depending on the microbiota and can be extended.
9. We have found that the single clindamycin IP regimen produces mild-to-moderate CDI symptoms when mice are challenged with 10^5 *C. difficile* spores of a ribotype 027 (RT027) strain. Severe disease is possible with increased doses or the incorporation of other factors, such as dietary manipulation. The *C. difficile* challenge dose should be optimized when using a new *C. difficile* or mouse strain, changing the antibiotic regime, or changing the source of the mice (e.g., a different commercial vendor).
10. The use of 1% BSA in final spore preparations and dilutions decreases clumping of spores, minimizes loss via adhesion to plasticware, and facilitates accurate enumeration [25].

11. The maximum dosing volume for mice is generally 10 mL/kg (check with your IACUC before proceeding). We generally challenge mice with 200 μL of spore solution. If younger or smaller mice are used, the volume may need to be reduced.
12. If new to oral gavage, it can be helpful to load only the required volume into the syringe. Experienced users may load enough volume to gavage a cage at a time, ensuring that each mouse receives the same dose.
13. Using a 1:10 dilution makes calculating the CFUs easier; however, depending on stool weight variability and number of samples, this can be time consuming and results in a low volume for low weight samples. An alternative method is to use a set volume of PBS (e.g., 250 μL).
14. Other common selective media include cycloserine–cefoxitin fructose agar (CCFA) [26], supplemented with taurocholic acid. CDMNT agar provides a lower background than CCFA in our hands.

References

1. Dong D, Su T, Chen W, Wang D, Xue Y, Lu Q et al (2023) Clostridioides difficile aggravates dextran sulfate solution (DSS)-induced colitis by shaping the gut microbiota and promoting neutrophil recruitment. Gut Microbes 15(1): 2192478. https://doi.org/10.1080/19490976.2023.2192478
2. Tomkovich S, Taylor A, King J, Colovas J, Bishop L, McBride K et al (2021) An osmotic laxative renders mice susceptible to prolonged clostridioides difficile colonization and hinders clearance. mSphere 6(5):e0062921. https://doi.org/10.1128/mSphere.00629-21
3. Lawley TD, Clare S, Walker AW, Goulding D, Stabler RA, Croucher N et al (2009) Antibiotic treatment of clostridium difficile carrier mice triggers a supershedder state, spore-mediated transmission, and severe disease in immunocompromised hosts. Infect Immun 77(9): 3661–3669. https://doi.org/10.1128/IAI.00558-09
4. Buffie CG, Jarchum I, Equinda M, Lipuma L, Gobourne A, Viale A et al (2012) Profound alterations of intestinal microbiota following a single dose of clindamycin results in sustained susceptibility to Clostridium difficile-induced colitis. Infect Immun 80(1):62–73. https://doi.org/10.1128/IAI.05496-11
5. Theriot CM, Koumpouras CC, Carlson PE, Bergin II, Aronoff DM, Young VB (2011) Cefoperazone-treated mice as an experimental platform to assess differential virulence of Clostridium difficile strains. Gut Microbes 2(6):326–334. https://doi.org/10.4161/gmic.19142
6. Chen X, Katchar K, Goldsmith JD, Nanthakumar N, Cheknis A, Gerding DN et al (2008) A mouse model of Clostridium difficile-associated disease. Gastroenterology 135(6):1984–1992. https://doi.org/10.1053/j.gastro.2008.09.002
7. Fachi JL, Felipe JS, Pral LP, da Silva BK, Correa RO, de Andrade MCP et al (2019) Butyrate protects mice from Clostridium difficile-induced colitis through an HIF-1-dependent mechanism. Cell Rep 27(3):750–61 e7. https://doi.org/10.1016/j.celrep.2019.03.054
8. Jenior ML, Leslie JL, Young VB, Schloss PD (2017) Clostridium difficile colonizes alternative nutrient niches during infection across distinct murine gut microbiomes. mSystems 2(4). https://doi.org/10.1128/mSystems.00063-17
9. Tomkovich S, Stough JMA, Bishop L, Schloss PD (2020) The initial gut microbiota and response to antibiotic perturbation influence Clostridioides difficile clearance in mice. mSphere 5(5). https://doi.org/10.1128/mSphere.00869-20
10. Onderdonk AB, Cisneros RL, Bartlett JG (1980) Clostridium difficile in gnotobiotic mice. Infect Immun 28(1):277–282. https://doi.org/10.1128/iai.28.1.277-282.1980

11. Reeves AE, Koenigsknecht MJ, Bergin IL, Young VB (2012) Suppression of Clostridium difficile in the gastrointestinal tracts of germ-free mice inoculated with a murine isolate from the family Lachnospiraceae. Infect Immun 80(11):3786–3794. https://doi.org/10.1128/IAI.00647-12
12. Collins J, Auchtung JM, Schaefer L, Eaton KA, Britton RA (2015) Humanized microbiota mice as a model of recurrent Clostridium difficile disease. Microbiome 3:35. https://doi.org/10.1186/s40168-015-0097-2
13. Auchtung JM, Preisner EC, Collins J, Lerma AI, Britton RA (2020) Identification of simplified microbial communities that inhibit Clostridioides difficile infection through dilution/extinction. mSphere 5(4). https://doi.org/10.1128/mSphere.00387-20
14. Tejada JN, Walters WA, Wang Y, Kordahi M, Chassaing B, Pickard J et al (2024) Prevention and cure of murine C. difficile infection by a Lachnospiraceae strain. Gut Microbes 16(1):2392872. https://doi.org/10.1080/19490976.2024.2392872
15. Chua M, Collins J (2024) Mice colonized with the defined microbial community OMM19.1 are susceptible to Clostridioides difficile infection without prior antibiotic treatment. mSphere 9(11):e0071824. https://doi.org/10.1128/msphere.00718-24
16. Studer N, Desharnais L, Beutler M, Brugiroux S, Terrazos MA, Menin L et al (2016) Functional intestinal bile acid 7alpha-dehydroxylation by Clostridium scindens associated with protection from Clostridium difficile infection in a gnotobiotic mouse model. Front Cell Infect Microbiol 6:191. https://doi.org/10.3389/fcimb.2016.00191
17. Aspinall ST, Hutchinson DN (1992) New selective medium for isolating Clostridium difficile from faeces. J Clin Pathol 45(9):812–814. https://doi.org/10.1136/jcp.45.9.812
18. Choo JM, Rogers GB (2021) Gut microbiota transplantation for colonization of germ-free mice. STAR Protoc 2(3):100610. https://doi.org/10.1016/j.xpro.2021.100610
19. Basic M, Bolsega S, Smoczek A, Glasner J, Hiergeist A, Eberl C et al (2021) Monitoring and contamination incidence of gnotobiotic experiments performed in microisolator cages. Int J Med Microbiol 311(3):151482. https://doi.org/10.1016/j.ijmm.2021.151482
20. Shelby RD, Tengberg N, Conces M, Olson JK, Navarro JB, Bailey MT et al (2020) Development of a standardized scoring system to assess a murine model of Clostridium difficile colitis. J Investig Surg 33(10):887–895. https://doi.org/10.1080/08941939.2019.1571129
21. Treuting PM, Snyder JM (2015) Mouse necropsy. Curr Protoc Mouse Biol 5(3):223–233. https://doi.org/10.1002/9780470942390.mo140296
22. Afrizal A, Jennings SAV, Hitch TCA, Riedel T, Basic M, Panyot A et al (2022) Enhanced cultured diversity of the mouse gut microbiota enables custom-made synthetic communities. Cell Host Microbe 30(11):1630-45e25. https://doi.org/10.1016/j.chom.2022.09.011
23. Etienne-Mesmin L, Chassaing B, Adekunle O, Mattei LM, Bushman FD, Gewirtz AT (2018) Toxin-positive Clostridium difficile latently infect mouse colonies and protect against highly pathogenic C. difficile. Gut 67(5):860–871. https://doi.org/10.1136/gutjnl-2016-313510
24. Lesniak NA, Schubert AM, Sinani H, Schloss PD (2021) Clearance of Clostridioides difficile colonization is associated with antibiotic-specific bacterial changes. mSphere 6(3). https://doi.org/10.1128/mSphere.01238-20
25. Edwards AN, Nawrocki KL, McBride SM (2014) Conserved oligopeptide permeases modulate sporulation initiation in Clostridium difficile. Infect Immun 82(10):4276–4291. https://doi.org/10.1128/iai.02323-14
26. George WL, Sutter VL, Citron D, Finegold SM (1979) Selective and differential medium for isolation of Clostridium difficile. J Clin Microbiol 9(2):214–219. https://doi.org/10.1128/jcm.9.2.214-219.1979

Chapter 15

Visualizing *C. difficile* During Murine Infection

Nicholas V. DiBenedetto and Aimee Shen

Abstract

Clostridioides difficile is a leading cause of healthcare-associated infections and displays phenotypic heterogeneity in key processes such as toxin gene expression (virulence) and sporulation (transmission). Understanding the mechanisms that regulate *C. difficile*'s physiology and pathogenesis requires tools capable of resolving phenotypically distinct subpopulations in vivo. In this chapter, we present fluorescence-based reporter systems for visualizing *C. difficile* gene expression in situ at single-cell resolution during infection. These spectrally compatible reporters are broadly applicable for investigating the spatial and temporal dynamics of gene expression for any target gene within the context of infection. Furthermore, they can be used in competition experiments between different *C. difficile* strains or to visualize *C. difficile* alongside other gut bacterial species, enabling spatial analysis of microbe–microbe interactions.

Key words Fluorescence, Expression, In vivo, Intermicrobial interactions

1 Introduction

Localizing bacteria within their native host environment has provided critical insights into host-microbe interactions and bacterial behavior in complex biological systems [1–5]. While techniques such as fluorescence in situ hybridization (FISH) have revealed the spatial distribution of bacterial species within tissues, using RNA-FISH to resolve the spatial organization of phenotypically distinct subpopulations within a single species can be technically challenging and expensive [6–8]. Here, we present a method for using chromosomally encoded fluorescent reporter strains to visualize in situ the location and magnitude of gene expression at the single-cell level, offering a more reliable, less labor-intensive, and cheaper alternative to methods such as RNA-FISH [9–12].

This chapter outlines methods for constructing *C. difficile* strains carrying chromosomally encoded fluorescent reporter constructs, including reporter design, mouse infection, tissue processing, microscopy, and image analysis of infected mouse tissue. We have identified constitutive reporters that allow all *C. difficile* cells

Kurt H. Piepenbrink and Jennifer M. Auchtung (eds.), *Clostridioides difficile: Methods and Protocols*, Methods in Molecular Biology, vol. 3046, https://doi.org/10.1007/978-1-0716-5328-9_15,

to be visualized alongside a gene-specific reporter, such as a toxin gene reporter [11, 12]. We have shown that single and dual-reporter strains yield reproducible infection phenotypes, from inducing consistent weight loss in murine models and allowing for the reliable detection of the reporter strains in colonic tissue sections [12]. Particular attention was given to optimizing the constitutive reporter to avoid fitness defects associated with fluorescent protein overexpression [12, 13]. Additionally, we confirmed that mNeonGreen and mScarlet-I3 are spectrally compatible reporters, allowing for clear differentiation of both signals in infected tissues [12, 14, 15]. While our work has primarily focused on visualizing toxin gene expression, this system is broadly applicable for studying the in situ expression of any gene of interest during infection, provided it is expressed at high enough levels to be detected at the single-cell level.

Since we also identified spectrally compatible constitutive *mScarlet-I3* and *mNeonGreen* fluorescent reporters that result in strains with comparable fitness, the methods described below can be used for competition assays to assess the role of a given gene in regulating the spatial distribution of *C. difficile* in the gut environment [12–15]. For instance, a wild-type strain carrying a chromosomally encoded constitutive *mScarlet-I3* reporter construct can be competed against a mutant strain carrying a chromosomally encoded constitutive *mNeonGreen* construct to determine whether a gene of interest alters the spatial distribution of *C. difficile* within the gut [12]. These constitutive reporter strains could also be used to compare the localization of *C. difficile* with another bacterium of interest that either expresses a constitutive fluorescent reporter construct or can be localized using immunofluorescence.

Finally, we have shown that this system provides sufficient resolution to detect abnormal *C. difficile* cell morphologies, since we have determined that a toxin gene overexpressing strain generates filamentous, elongated, and curved cells specifically during the acute phase of infection [12]. Taken together, the development of these tools provides an effective and reliable methodology for visualizing multiple *C. difficile* populations during infection, offering a robust platform for investigating gene function, phenotypic heterogeneity, spatial dynamics in vivo, and microbe–microbe as well as host–microbe interactions.

1.1 Considerations When Designing Fluorescent Reporter Strains

While this chapter does not provide detailed methods pertaining to the genetics required to construct fluorescent reporter strains, it is important to note that two well-established systems are commonly used to genetically manipulate laboratory strains of *Clostridioides difficile*. The *pyrE* allelic exchange system is often used with the 630Δ*erm*Δ*pyrE* strain [16], while a CRISPR-based system is more commonly applied to the more virulent R20291 strain because of the higher efficiency of this method [17, 18]. For the purposes of this chapter, the choice of system is less important than selecting the most efficient method for the specific strain or experimental use case.

When designing a dual fluorescent reporter construct, several factors should be considered to ensure optimal visualization. One must determine which fluorescent reporter gene will be constitutively expressed and which will be driven by the promoter of the gene of interest. A key consideration is the expected expression level of the target gene. *mScarlet-I3* offers slightly higher sensitivity than *mNeonGreen* due to the higher autofluorescence of colonic content in the green channel [19, 20], making it the preferred reporter for genes with low expression levels due to its improved signal-to-noise ratio. For strongly expressed target genes, such as toxin and certain sporulation genes, either reporter is suitable.

For constitutive reporters, we tested multiple combinations of promoters and fluorescent proteins to optimize single-cell brightness. The optimal constructs used the P_{cwp2} promoter driving *mScarlet-I3* expression and the P_{slpA} promoter driving *mNeonGreen* expression. Both *cwp2* and *slpA* encode cell wall proteins that are highly expressed in *C. difficile*, providing strong and constitutive fluorescence signals [11, 12, 21, 22]. To maximize the levels of mNeonGreen and mScarlet, we also codon-optimized the *mNeonGreen* and *mScarlet-I3* genes for *C. difficile*.

In our case, we predominantly work with the 630 strain, so we constructed our fluorescent reporters using the *pyrE* allele-coupled exchange system, which allows us to move two fluorescent constructs onto the chromosome [11, 12, 16]. We first identified a neutral locus downstream of the sporulation gene *sipL* [23] and used allelic exchange to integrate our constitutive P_{cwp2}*::LP-mScI3* reporter into 630Δ*erm*Δ*pyrE*. For our gene-specific reporter, *the tcdA* toxin gene reporter, we used the pMTL-YN1C plasmid to insert P_{tcdA}*::mNG* into the *pyrE* locus of 630Δ*erm*Δ*pyrE* [16]. This strategy allowed us to create a dual-reporter strain that constitutively expresses *mScarlet-I3* and induces *mNeonGreen* expression in response to the activation of the P_{tcdA} promoter [12]. While the example described here uses the *tcdA* promoter, it can easily be replaced with the promoter of any gene of interest to analyze the expression levels and dynamics of the gene at the single-cell level.

After constructing a fluorescent reporter strain, it is critical to ensure that the constructs do not cause a growth defect using growth curves and that they produce fluorescence consistent with the expected expression patterns for the gene of interest using microscopy. While these analyses are performed during growth in vitro, these validation steps are essential to ensure that the reporter strain is functioning as intended before proceeding to downstream in vivo applications. As described below, the effect of the constructs on the fitness of *C. difficile* must also be assessed during murine infection using competitive infections (*see* **Note 1**).

2 Materials

2.1 Mouse Infection and Tissue Collection

All materials should be sterilized using an autoclave before use

1. 7-week-old female C57BL/6 mice from Jackson Laboratories (ideally, at least 4 mice per condition per experiment, across 2–3 replicates to achieve 8–12 mice total).
2. Purified *C. difficile* spores with fluorescent reporter constructs integrated onto the chromosome.
3. Diet: Teklad irradiated global 18% protein rodent diet 2918 (Inotiv).
4. Water.
5. Large sterile rat cage (for microbiota normalization).
6. Bedding and nestlets.
7. Sterile water bottles.
8. Mouse cages for housing infected mice.
9. Large forceps for handling mice (use a separate pair for each *C. difficile* strain to avoid cross-contamination) (Amazon; Liuna 304 Stainless Steel Tweezers 12 inch long).
10. Tri-pour 1-L plastic cups for weighing mice (use a separate cup for each *C. difficile* strain to avoid cross-contamination) (Consolidated Plastics; #45321).
11. Digital scale.
12. 1-mL syringe (Fisher Scientific; 14-823-30).
13. Autoclavable gavage needles (Cadence Science; #7921).
14. 1.5-mL Eppendorf tubes (for spore inoculum and tissue collection).
15. Eppendorf tube rack.
16. Cefoperazone (Research Products International; #C57600-5.0).
17. Clindamycin (Sigma-Aldrich; #C5269-50MG).
18. 1x PBS (phosphate-buffered saline) (Fisher Scientific; #BP3991).
19. 4% PFA in 1x PBS (Electron Microscopy Sciences; #15710).
20. Aluminum foil.
21. Scissors (Cole-Parmer; #EW-10818-00).
22. Fine-tip forceps (Fisher Scientific; #50-242-49).
23. 100% ethanol (Fisher Scientific; #BP2818-500).
24. Dry ice.

25. Seal' N Freeze Box (Electron Microscopy Sciences; #62642-03).
26. Tissue cassettes (Electron Microscopy Sciences; #62642-01).
27. CryoTek OCT (Sakura; #4583).
28. Alcohol-resistant marker (Avantor; #807656004495).
29. Razor blade.
30. Petri dish.

2.2 Staining Tissue Sections for Microscopy

1. Large tissue culture dish (Alkali Scientific; #PD3500).
2. KimWipes (Fisher Scientific; #06-666).
3. Distilled water.
4. Samples (tissue slides stored at −80 °C).
5. Bovine Serum Albumin (BSA) (Fisher Scientific; PI23210).
6. Triton X-100 (Sigma-Aldrich; X100-5ML).
7. 1x PBS (phosphate-buffered saline) (Fisher Scientific; #BP3991).
8. 0.22-μm sterile filter (for buffer sterilization) (Fisher Scientific; #09-761-102).
9. Hydrophobic barrier pen (Vector Laboratories; #H-4000).
10. DAPI (4′,6-diamidino-2-phenylindole) (Sigma-Aldrich; D9542-10MG).
11. Phalloidin 647 (Santa Cruz Biotechnology; #sc-363797).
12. P1000 pipette and pipette tips.
13. 22 × 50 mm coverslips (Electron Microscopy Sciences; #72200-40).
14. 25 × 75 × 1 mm slides (Fisher Scientific; #12-550-143).
15. ProLong Diamond Antifade mounting medium (Fisher Scientific; #P36965).
16. Light-protected container or box (for incubation, can place a lid of a Styrofoam box over the tissue culture dish and slides to block light).

2.3 Microscopy of C. difficile-Infected Tissue Sections

1. Leica DMi8 Thunder microscope.
2. HC PL APO 20×/0.80 dry objective (Leica).
3. Leica DFC9000 GTC sCMOS camera.
4. Lumencor Spectra-X LED light source.
5. Leica XLED-QP quadruple-band dichroic mirror.
6. External filter wheel (Leica).
7. Fluorescence filter sets (see table above).

8. LAS X software with Navigator and Thunder deconvolution modules.
9. Mounted and stained tissue slides (10-μm OCT sections with DAPI, mScarlet-I3, mNeonGreen, Phalloidin).

3 Methods

3.1 Mouse Infection and Tissue Collection

Day 0–10: Microbiota Normalization

1. House cage-mate mice together in a large sterile rat cage for 10 days (Fig. 1).
   - *Per the Guide for the Care and Use of Laboratory Animals, there is a minimum floor space required per animal (>25 g = 97 cm^2); therefore, most standard mouse cages can house up to 5 mice, while rat cages can hold ~20 mice. It will vary depending on the cage sizes at your institution* [24].
2. Provide irradiated Lab Diet 2918 and sterile drinking water ad libitum throughout this period.

Day 10–20: Antibiotic Pretreatment with Cefoperazone [25]

3. Add cefoperazone to the drinking water at a final concentration of 0.5 mg/mL.
4. Continue cefoperazone treatment in the drinking water ad libitum for 10 days.
   - Cefoperazone is light sensitive, so it is recommended to replace cefoperazone every 2–3 days to mitigate antibiotic degradation.

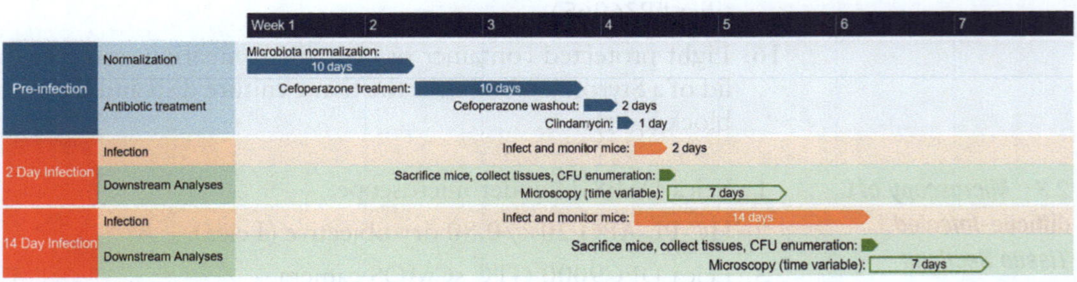

Fig. 1 Timeline of murine infection. Depending on when the mice are euthanized, the murine infections can take between 6 and 8 weeks to complete, if the mice will be euthanized at 2 days, or 14 days, respectively. There are three parts to the infection: preinfection, which includes normalization of the microbiota upon arrival to the animal facility, followed by antibiotic treatment with a 10-day course of cefoperazone at 0.5 mg/mL and then a single IP injection with clindamycin at 1 mg/kg. After the antibiotic treatment, mice are infected with 1×10^6 spores and euthanized at either 2 or 14 days postinfection. Tissues are collected at the time of euthanasia for CFU enumeration and microscopy analyses

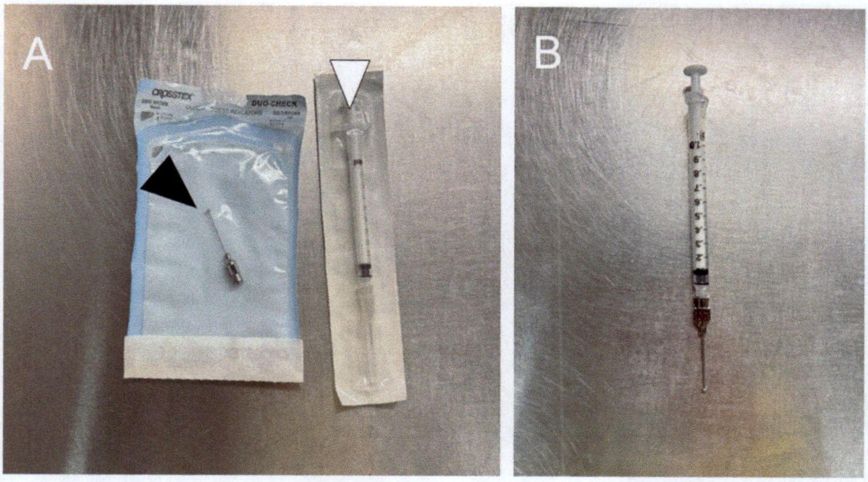

Fig. 2 Equipment for infecting mice with *C. difficile* spores. Reusable metal gavage needles (*black arrow*) are autoclaved to ensure sterility and are used with 1-mL syringes (*white arrow*) (**a**). To infect mice, the gavage needle is removed from the sterile pouch and attached to a 1-mL syringe (**b**). Carrier volumes for infection are 200 μL, so one syringe can be used to infect 5 mice, which is the maximum number of mice allowed per cage

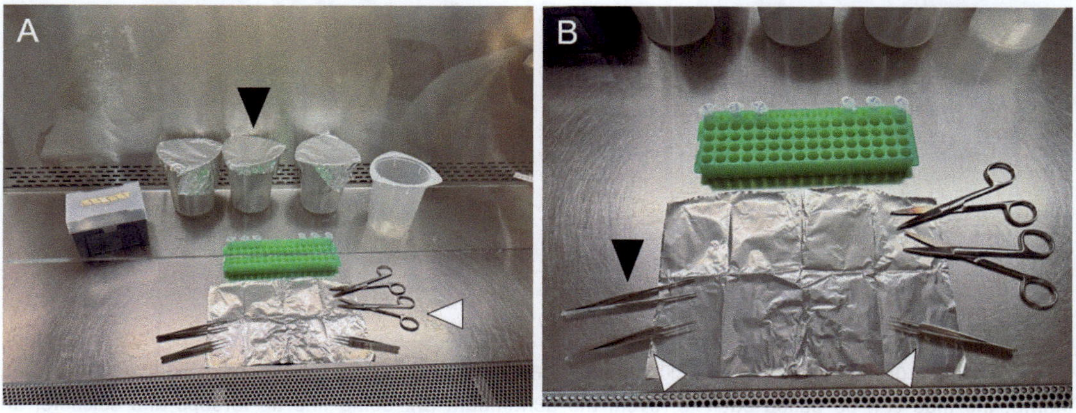

Fig. 3 Equipment for sacrificing mice and collecting tissues. Autoclaved aluminum foil is used as a sterile field to harvest tissues on. Plastic cups are filled with scissors and forceps and are covered with aluminum foil and autoclaved to ensure sterility (*black arrow*). An empty plastic cup is used to place dirty tools once tissue collection is complete. An Eppendorf tube rack is present to collect tissues, and a sterile field with scissors and forceps is prepared for mouse tissue harvest (*white arrow*) (**a**). A zoomed-in view of the sterile field where tissues will be collected from mice, showing a tube rack, two scissors (one for the mouse hide, and one for cutting tissues), blunt-tip forceps (*black arrow*) for holding the mouse hide, and two fine-tip forceps (*white arrows*) for handling tissues (**b**)

Day 20–22: Washout Period

5. Replace the antibiotic water with sterile drinking water for a 2-day washout period.

Day 22: Clindamycin Injection

6. Prepare clindamycin at 1 μg/μL in 1x PBS and filter-sterilize.

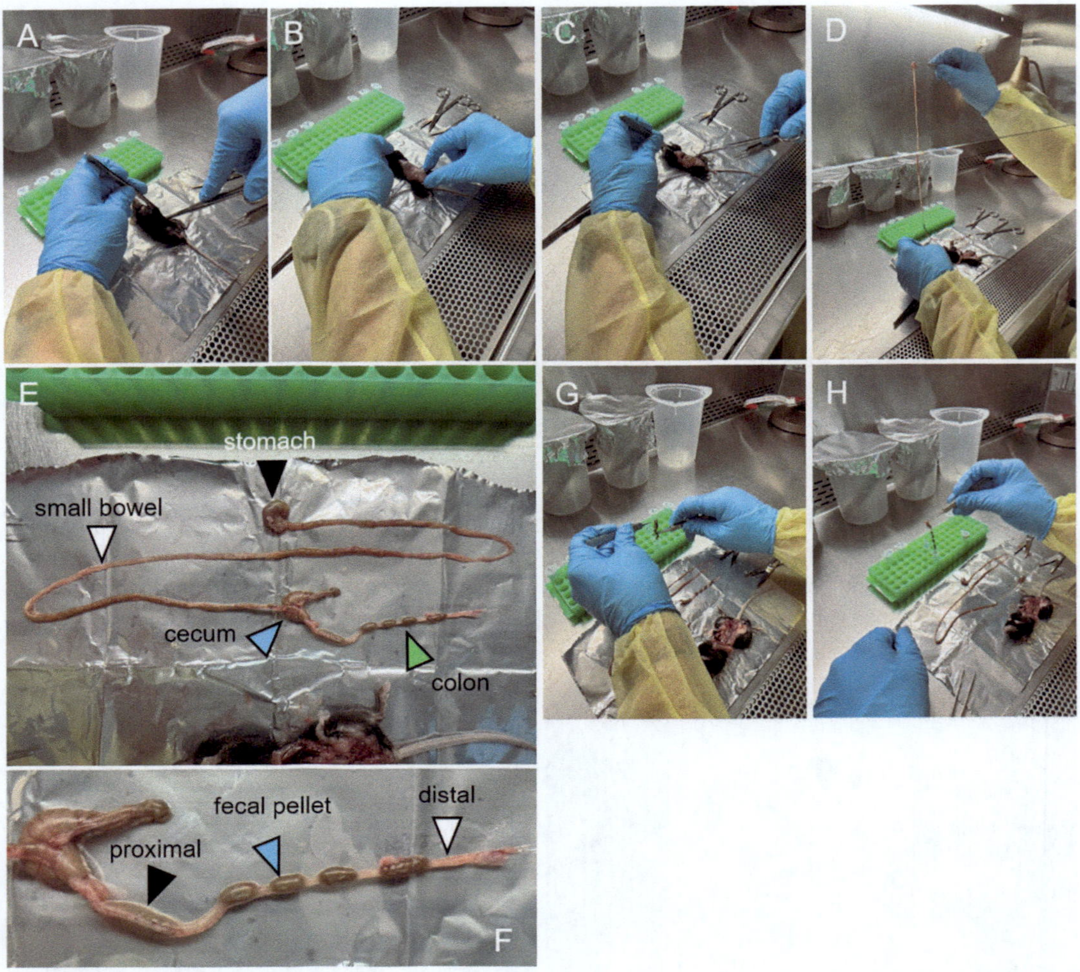

Fig. 4 Harvesting tissues from mice. Blunt-tip forceps and a pair of scissors are used to cut through the hide on the abdomen of the mouse (**a**). By pinching the hide on each side of the incision with the index finger and thumb, the hide can be pulled back to reveal the peritoneum (**b**). Using fine-tip forceps and scissors, an incision can be made in the peritoneum and further opened to reveal the gastrointestinal (GI) tract (**c**). A cut is made at the junction between the esophagus and the stomach and at the anus, which allows for the removal of the GI tract (**d**). The entire GI tract can be laid out on the sterile field including the stomach (*black arrow*), the small bowel (*white arrow*), the cecum (*blue arrow*), and colon (*green arrow*) (**e**). A zoomed-in image of the cecum and colon, showing the proximal colon (*black arrow*), distal colon (*white arrow*), and fecal pellets (*blue arrow*) (**f**). A cut at the base of the cecum allows the cecal content to be harvested by squeezing the cecum with fine-tip forceps (**g**). The colon can then be placed inside 4% paraformaldehyde for fixation (**h**)

7. Weigh each mouse by placing it in a sterile Nalgene cup using large forceps, and then measure weight on a digital scale inside a biosafety cabinet.
8. Inject each mouse intraperitoneally (i.p.) with clindamycin so that each receives 10 μg per gram of body weight.

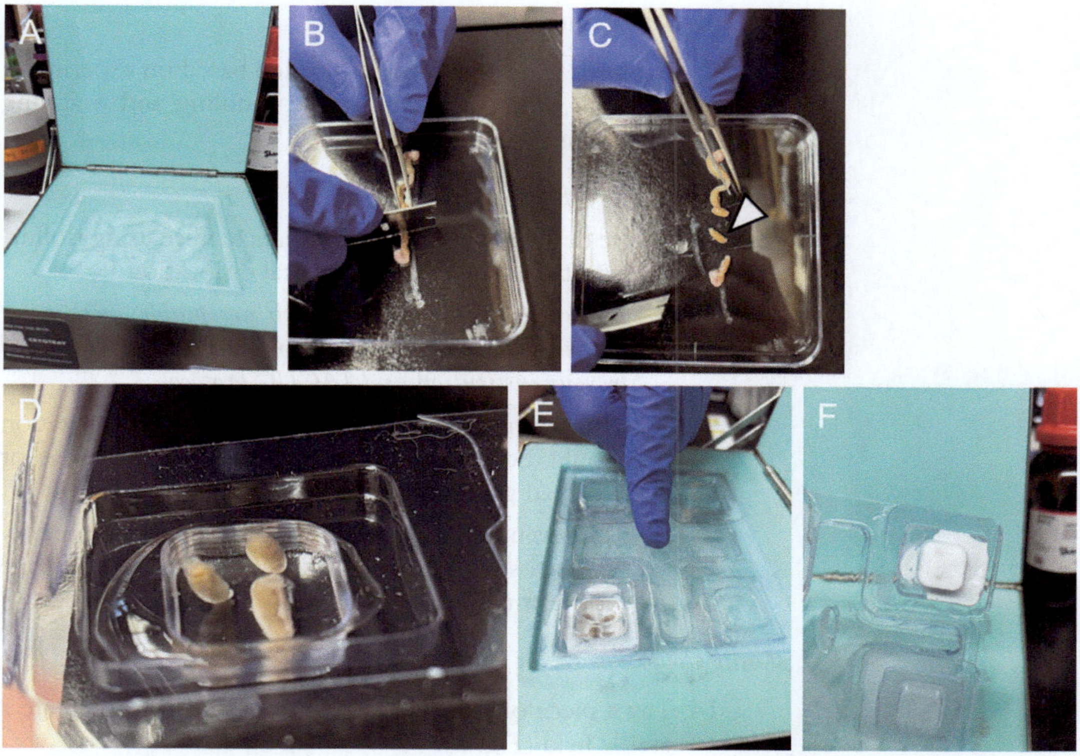

Fig. 5 Tissue cross-sectioning and OCT embedding. Dry ice is placed inside of the cryotray, and 100% ethanol is added to make an ethanol bath that will be used to freeze the tissues in OCT (**a**). Colon tissue is placed into a sterile square petri dish. Using a razor blade and fine-tip forceps, cuts are made such that the tissue contains a fecal pellet. Sectioning tissues through a fecal pellet increases the probability that *C. difficile* will be present in the sections when imaging (**b**). A section of colonic tissue containing a fecal pellet (*white arrow*) (**c**). Three cross sections are made, and the tissue sections are placed in a tissue cassette containing OCT. The tissues are upright in the OCT, such that when they are cryosectioned, they will produce cross sections through the tissue (**d**). The tissue cassette is placed in the ethanol-dry ice bath for 30 seconds, allowing for the OCT to completely freeze (**e**, **f**)

Day 23: C. difficile Infection

9. Transfer mice to smaller sterile mouse cages, housing 4 mice per cage.
10. Inoculate each mouse via oral gavage with 1×10^6 *C. difficile* spores in 200 μL of sterile 1x PBS [26] (Fig. 2).
11. Record baseline weights by:
    - Transferring mice with large forceps into sterile plastic Nalgene cups.
    - Weighing them on a digital scale inside a biosafety cabinet.
    - Using separate forceps and Nalgene cups for mice infected with different *C. difficile* strains to prevent cross-contamination.

Postinfection Monitoring

12. Monitor and weigh mice as needed based on experimental goals, using the same sterile handling and weighing procedures.

Experimental Endpoint

13. At the designated endpoint, euthanize mice using isoflurane anesthesia followed by cervical dislocation.
    - *Isoflurane and CO_2 are both acceptable methods of euthanasia. It is important to follow what is approved by your institution and IACUC protocol.*
    - The timing of euthanasia and reporter visualization should be based on the experimental question or expected gene expression kinetics.
    - To prevent tissue drying and to maintain the integrity of the tissues, euthanize and collect tissue from one mouse at a time.

Tissue Collection

14. Lay a piece of sterilized aluminum foil down to create a sterile field (Fig. 3).
15. For each mouse, use 2 scissors and 2 forceps:
    - One pair of scissors to cut through the hide.
    - One pair to cut through the peritoneum and colon.
16. Using fine-tip forceps and scissors; remove the cecum, colon, and anus; and place them into an Eppendorf tube containing 1 mL of 4% PFA in 1 × PBS. Fix the tissues for 4 h at room temperature (Fig. 4).
    - Keep tissues connected to maintain the orientation of the proximal and distal colon.

Cross-Sectioning Tissues and OCT Embedding

17. Prepare a dry ice–ethanol bath by filling an ice bucket with dry ice and then adding 100% ethanol. Continue adding ethanol until the bubbling slows to a gentle simmer (Fig. 5).
18. Label tissue cassettes with an alcohol-resistant marker—one cassette per mouse.
19. Fill each cassette with OCT.
20. Remove the colon from fixative and place in a clean, sterile petri dish.

21. Use a razor blade to cut 0.5-cm cross sections of colon; ideally, the section will contain a fecal pellet. Place 2–3 cross sections into a single tissue cassette.
    - Retaining fecal pellets in the cross sections helps retain *C. difficile* in the sections while preserving the spatial context. Because *C. difficile* is nonadherent, it is easily washed away during fixation and sample washes.
22. Use fine-tip forceps to stand cross sections upright in the OCT within the cassette.
23. Close the cassette and place it into the dry ice–ethanol bath for 30 s. Ensure the OCT has fully solidified (Fig. 5e–f).
    - Avoid completely submerging the cassette in ethanol, as ethanol entry can create artifacts in the tissue (*see* **Note 2**).
24. Once the OCT is fully frozen, transfer the cassette to a −80 °C freezer until cryosectioning and slide mounting.

3.2 Staining Tissue Sections for Microscopy

While this protocol does not cover cryosectioning or slide mounting, it is important to note that 10-μm tissue sections are typically used. Three sections are mounted per slide, and three slides are prepared from each OCT-embedded tissue block. This approach provides multiple sections with intact tissue architecture for imaging. If a section is damaged or folded during mounting, other sections can be used instead. Because fluorescent reporters are prone to photobleaching, it is not ideal to image the same slide repeatedly. If additional imaging is needed, use a different section on the slide that has not yet been exposed. If your lab lacks a cryotome, a histology core facility can perform the sectioning and mounting.

*For this protocol, applying and removing buffers and stains should be done with a pipette as gently as possible to minimize perturbing the tissues (*see* **Note 3**). Carefully tilt the slide at an angle so that the buffer pools towards the edge of the slide, which allows for the more gentle removal of the liquid (Fig. 6).*

Prepare a humidified chamber

1. Line the bottom of a large tissue culture dish covered in aluminum foil with KimWipes dampened with distilled water. This chamber will prevent evaporation during staining steps (Fig. 6).

Equilibrate slides

2. Remove frozen tissue slides from −80 °C and allow them to equilibrate at room temperature for 30 min.

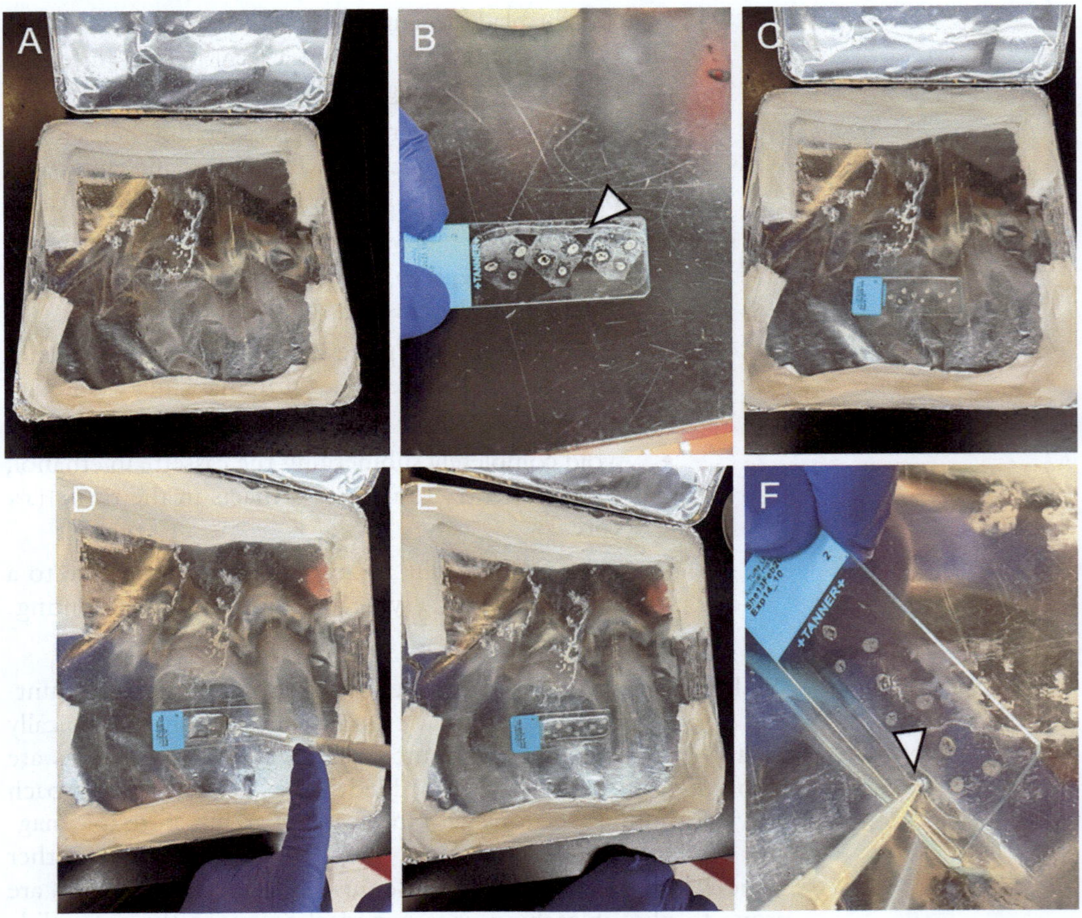

Fig. 6 Staining tissues for microscopy. A container made from a large square petri dish covered in aluminum foil is used to stain light-sensitive tissue samples. Large KimWipes are soaked with distilled water and placed around the edges to make a humidified chamber (**a**). Frozen slides are thawed at room temperature on the bench out of light for 30 min. After thawing, a hydrophobic marker is used to trace around the edges of the tissue sections (*white arrow*) to ensure that buffers do not run off the edge of the slide (**b**). The slide is placed in the humidified chamber, and buffer/stain is applied to the tissue sections so that all tissues are completely covered (**c**–**e**). To remove the buffer, the slide is tilted on edge so that the buffer builds up along the hydrophobic marker (*white arrow*). This allows for a more complete removal of buffer during wash steps (**f**)

Prepare blocking buffer

3. Prepare blocking buffer consisting of PBS with 3% BSA and 0.1% Triton X-100.
   - For 250 mL: dissolve 7.5 g BSA and add 250 μL 100× Triton X-100 in PBS.

Mark tissue area

4. Using a hydrophobic barrier pen, draw a border around each tissue section to contain reagents during staining.

Block

5. Apply ~250 μL of blocking buffer to each section. Incubate slides in the humidified chamber for 30 min at room temperature.
6. After 30 min, carefully remove the blocking buffer by pipetting with a P1000.

Primary staining with DAPI and Phalloidin

7. Prepare staining solution in blocking buffer with DAPI (1: 500) and Phalloidin 647 (1:100).
   - For 3 mL: add 6 μL DAPI and 30 μL Phalloidin 647 to 2964 μL blocking buffer.

     Incubate tissue sections with staining solution for 45 minutes at room temperature, protected from light.
   - *Three milliliters of stain should cover ~12 slides, as each slide needs ~250 μL to be completely covered.*

Wash

8. Wash sections three times for 5 min each in fresh blocking buffer.

Mount

9. Apply two drops of ProLong Diamond Antifade mounting medium to each section. Carefully apply a coverslip at an angle, using gentle pressure to avoid air bubbles.

Cure

10. Incubate slides at room temperature overnight in the dark to allow the mounting medium to fully cure and fluorophores to mature.

3.3 Microscopy of C. difficile-Infected Tissue Sections

*The microscopy protocol will vary depending on the brand and model of microscope; however, the protocol will generally be the same. For best results, use equivalent or similar hardware and software. *See* **Note 4** for additional suggestions if you encounter difficulties identifying fields with *C. difficile.**

All images were acquired using a Leica DMi8 Thunder widefield microscope equipped with an HC PL APO 20x/0.80 dry objective and a Leica DFC9000 GTC sCMOS camera. Excitation was provided by a Lumencor Spectra-X multi-LED light source with integrated excitation filters. All fluorescent channels were imaged using a quadruple-band XLED-QP dichroic beam-splitter and external filter wheel (Leica).

Imaging was performed with the filter and acquisition settings described in Table 1.

Table 1
Summary of microscope excitation and emission settings

Channel	Excitation (nm)	Intensity	Exposure (ms)	Dichroic mirror (nm)	Emission filter (nm)
mScarlet-I3	550	33%	200	570	595/40 (Leica)
mNeonGreen	470	100%	300	419	515/40 (Leica)
DAPI	395	10%	50	415	430/36 (Leica)
Phalloidin-647	640	100%	300	660	720/50 (Leica)

Create a New Project

1. Open LAS X software.
2. Start a new project.
3. Select the correct settings and configuration for imaging based on Table 1.

Set Image Orientation

4. In the Configuration screen, enable Vertical Flip under "Image Flip."

Select Tissue in LAS X Navigator

5. Open the Navigator (grid icon).
6. Select the DAPI channel, and then click Live.
7. Use the Z-course tool to locate the DAPI signal plane.
8. Use spiral scanning to preview the entire tissue.
9. Use the circle tool to define the region to be imaged (usually 20–30 tiles for a full cross section).

Set Z-Stack Parameters

10. Go to the Image tab and click Z → Z-stack.
11. Focus just above the tissue in the DAPI channel and click Begin.
12. Scroll through the tissue to just below and click End.
13. Set Z-step size to 5 μm.
    - *While 5-μm step sizes may seem too large to capture fine details, it will be sufficient for visualizing C. difficile and spatial data. Smaller step sizes require more imaging and ultimately larger file sizes. This results in longer image acquisition and postprocessing time as well as specialized storage solutions for larger files* (e.g., *~50 GB* vs. *5–10 GB). Smaller Z-step sizes should only be used if absolutely necessary.*

Configure Thunder Settings

14. Enable Instant Computational Clearing (ICC) (lightning bolt icon).
    - *Leica's Instant Computational Clearing (ICC) is a real-time image processing technique built into Leica's THUNDER Imagers. It is designed to improve image clarity in wide-field fluorescence microscopy by removing out-of-focus background signal* [27].
15. In Advanced Settings, set the Feature Scale for:
    - mNeonGreen and mScarlet-I3 = 2500 nm,
    - *This is roughly half the length of a C. difficile cell and helps the software differentiate real fluorescent signals from background noise.*

Acquire Images

16. Stop Live preview.
17. Double-check all settings.
18. Click Start to begin acquisition (typically 5–10 min, ~10–20 GB).
19. During acquisition, verify that the Z-stack fully covers the tissue for the first few tiles.

Merge Mosaic

20. In Tilescan ICC, click Mosaic Merge. This process will stitch all of the tiles together into one large image.
21. Press Merge (bottom right of the window).

Process Z-Stacks

22. Close the Navigator window.
23. Go to the Process tab.
24. Under Open Projects, select your tile scan (ICC).
25. Go to Process Tools → Projection.
    - *Maximum Projection is an image processing method that flattens a Z-stack into a single 2D image by displaying only the brightest pixel at each (x, y) position across all focal planes. This highlights the most prominent fluorescent signals throughout the tissue depth* [28].
26. Set Method to Maximum and click Apply.

Save Final Files

27. Save the raw file as RAW.

28. Drag the ICC_merged_processed file into a new project window and save as [filename]_Processed.
    - *This step is necessary so that "working" files can be stored on smaller storage capacity portable drives for easy editing, while larger raw files can be stored on larger storage capacity drives.*
29. Delete intermediate merged files to conserve storage.

3.4 Adjusting Images in FIJI for Presentation and Data Generation

1. Open up Image J and then drag your.lif file that contains your images of mouse tissue sections into the task bar.
2. Select images that you want to open. The.lif file will contain any images saved in a single project.
3. In the task bar, open up the brightness and contrast tools as well as the channel tool.
4. Scroll through the different channels and adjust the brightness and contrast of in each channel to best visualize tissues and *C. difficile* (Fig. 7).
    - It is important to keep brightness and contrast settings the same across multiple images if they will be compared to each other. In doing so, you will not artificially skew the brightness of the reporters.
5. To gather fluorescent intensity information, use the line tool to draw a line from pole to pole of *C. difficile* cells.
6. Using the line tool, you will be able to obtain information such as fluorescent intensity across all channels (mean or median), as well as cell length and curvature.

4 Notes and Troubleshooting

1. *Competition Assays*

   When co-infecting mice with two different fluorescent reporter strains, it is essential to quantify the input spore ratio of each strain. While the goal is to inoculate with equal numbers of spores from each strain, this is not always achieved precisely. To determine the actual input ratio, plate the inoculum on BHIS supplemented with taurocholate to promote spore germination. Once colonies have grown, pick individual colonies and resuspend each in 100 μL of PBS in a 96-well plate. Measure fluorescence using a plate reader to quantify mScarlet signal, allowing you to calculate the ratio of mScarlet to mNeonGreen spores in the inoculum.

   It is also possible to use a fluorescence scanner to detect the mScarlet signal in colonies on a plate. For example, a BioRad ChemiDoc can be used to visualize fluorescent colonies on agar plates.

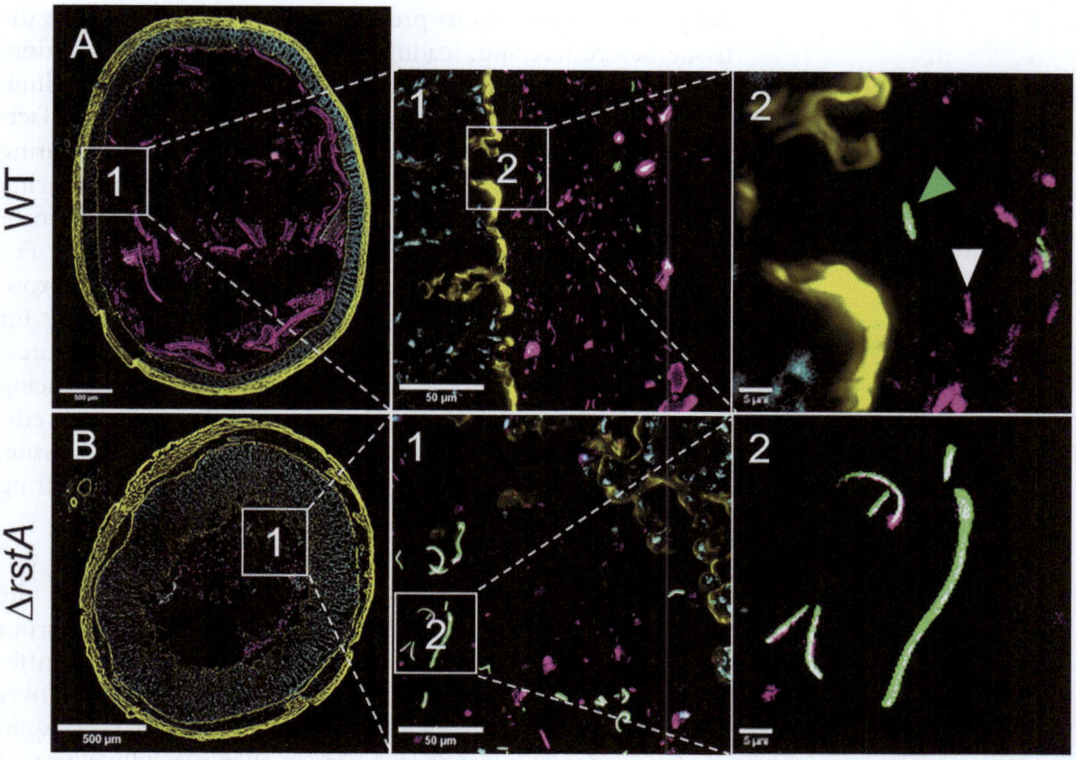

Fig. 7 Images of mouse colonic cross sections infected with fluorescent *C. difficile* reporter strains. Images of mouse colonic tissue sections showing the epithelium (*yellow, F-actin*) and epithelial nuclei (*cyan, DAPI*). All *C. difficile* cells fluoresce mScarletI3 (*white arrow, magenta,* P_{cwp2}*::LP-mScl3*), while the cells expressing the toxin gene *tcdA* fluoresce mNeonGreen (*green arrow, green,* P_{tcdA}*::mNG*) (**a**, **b**). Mice infected with wild-type *C. difficile* heterogeneously express *tcdA* (**a**). Mice infected with a toxin overexpression mutant Δ*rstA* show increased frequency of cells that are Toxin-ON, as well as a unique elongated and curved morphology (**b**)

Note: Due to background autofluorescence in the green channel, unmarked strains cannot be reliably distinguished from mNeonGreen. Therefore, when competing a reporter strain against an unmarked strain, it is best to use mScarlet as the reporter.

To assess the relative abundance of each strain after infection, collect fecal material from the mice, plate it on BHIS with taurocholate, and isolate single colonies as described above. Measure fluorescence to determine the output ratio of each strain. The competitive index (CI) can then be calculated by comparing the output ratio to the input ratio:

$$\begin{aligned}\mathrm{CI} = &\left(\mathrm{mScarlet_{output}}/\mathrm{mNeonGreen_{output}}\right)\\ &\div \left(\mathrm{mScarlet_{input}}/\mathrm{mNeonGreen_{input}}\right)\end{aligned}$$

2. *OCT Embedding Tissue Sections*

If ethanol contaminates OCT blocks during the flash freezing of mouse colonic tissue sections, it can negatively impact tissue quality and downstream analyses. Ethanol is a dehydrating agent, so its presence can partially dehydrate the tissue before freezing, leading to tissue shrinkage, distortion, or collapse of delicate structures such as the colonic epithelium [29, 30]. This often results in visible morphological artifacts like loss of crypt architecture, wrinkling, or tearing during cryosectioning. Additionally, ethanol lowers the freezing point of the embedding medium, which can cause uneven or brittle freezing, leading to cracked or crumbling sections. For experiments involving fluorescence microscopy, ethanol exposure can quench or alter fluorescent signals, particularly for sensitive fluorophores like GFP derivatives [31–33]. Fluorescent proteins may become denatured or aggregated, reducing signal intensity and increasing background. Furthermore, ethanol can prevent proper infiltration of OCT into the tissue, leading to separation artifacts or ice crystal formation during freezing [30].

3. *Staining Slides*

When washing slides, avoid pipetting directly over the tissue sections, as this can dislodge fecal material and disrupt the spatial distribution of bacteria. To prevent this, apply buffer or stain to the edges of the slide and allow it to gently flow over the tissue. When removing liquid, tilt the slide so that the liquid pools against the hydrophobic barrier, then carefully pipette it off without disturbing the sample.

When applying ProLong Diamond antifade mounting medium, small bubbles can form within the tissue sections. To minimize this, centrifuge the bottle briefly before use. It is best to place a drop of antifade directly over the tissue, as it may not spread evenly through the fecal material once the coverslip is applied. Always store ProLong Diamond at 4 °C to prevent premature curing. If excessive bubbles appear, the reagent may be old and should be replaced. Do not press on the coverslip to remove bubbles, as this can shift the fecal material and bacteria, compromising the spatial organization of cells in the tissue.

4. *Microscopy of Tissue Sections*

When searching for *C. difficile* cells in colonic tissue sections, areas containing fecal material should be prioritized, as the fecal pellet helps prevent bacteria from washing away during processing. If no cells are detected despite the presence of a fecal pellet, it is advisable to examine a different cross section. This may be due to the fact that *C. difficile* occupies distinct niches within the colon, and some areas may naturally lack bacterial cells, or the bacteria were washed away during the fixation, sectioning, or washing of the samples. Additionally, if working with a strain that has colonization defects, there may be regions of the colon entirely devoid of *C. difficile*.

Although the fluorescent reporters described in this chapter are stable and mature rapidly, they are still susceptible to photobleaching, which can significantly reduce signal intensity after imaging. If reimaging is necessary, it is best to use a different, previously un-imaged cross section from the same mouse to ensure strong fluorescence and reliable visualization of cells.

Data Availability

Plasmids generated to make the strains in this article were deposited with Addgene and are available under the corresponding accession numbers: pMTL-YN1C Promoter-less *mNeonGreen* (Addgene ID: 251223), pMTLYN1C P_{tcdA}::*mNeonGreen* (Addgene ID: 251224), pMTL-YN1C P_{slpA}::*mNeonGreen* (Addgene ID: 251225), and pMTL-YN1C P_{cwp2}::*LP-mScI3 (G228R)* (Addgene ID: 251226).

References

1. Tropini C, Earle KA, Huang KC, Sonnenburg JL (2017) The gut microbiome: connecting spatial organization to function. Cell Host Microbe 21:433–442
2. McCallum G, Tropini C (2024) The gut microbiota and its biogeography. Nat Rev Microbiol 22:105–118
3. Mondragón-Palomino O et al (2022) Three-dimensional imaging for the quantification of spatial patterns in microbiota of the intestinal mucosa. Proc Natl Acad Sci 119:e2118483119
4. Davis KM, Mohammadi S, Isberg RR (2015) Community behavior and spatial regulation within a bacterial microcolony in deep tissue sites serves to protect against host attack. Cell Host Microbe 17:21–31
5. Lötstedt B, Stražar M, Xavier R, Regev A, Vickovic S (2024) Spatial host–microbiome sequencing reveals niches in the mouse gut. Nat Biotechnol 42:1394–1403
6. Young AP, Jackson DJ, Wyeth RC (2020) A technical review and guide to RNA fluorescence in situ hybridization. PeerJ 8:e8806
7. Higo S, Ishii H, Ozawa H (2023) Recent advances in high-sensitivity in situ hybridization and costs and benefits to consider when employing these methods. Acta Histochem Cytochem 56:49–54
8. Cao Z et al (2023) Spatial profiling of microbial communities by sequential FISH with error-robust encoding. Nat Commun 14:1477
9. Uliczka F et al (2011) Monitoring of gene expression in bacteria during infections using an adaptable set of bioluminescent, fluorescent and colorigenic fusion vectors. PLoS One 6: e20425
10. Campbell-Valois F-X, Sansonetti PJ (2014) Tracking bacterial pathogens with genetically-encoded reporters. FEBS Lett 588:2428–2436
11. Donnelly ML et al (2022) Development of a dual-fluorescent-reporter system in Clostridioides difficile reveals a division of labor between virulence and transmission gene expression. mSphere 7:e0013222
12. DiBenedetto NV, Donnelly-Morell ML, Kumamoto CA, Shen A (2025) Visualizing phenotypic heterogeneity and single-cell morphology in situ during gut infection. 2025.04.30.651596 Preprint at https://doi.org/10.1101/2025.04.30.651596
13. Kafri M, Metzl-Raz E, Jona G, Barkai N (2015) The cost of protein production. Cell Rep 14:22–31
14. Shaner NC et al (2013) A bright monomeric green fluorescent protein derived from Branchiostoma lanceolatum. Nat Methods 10:407–409
15. Gadella TWJ et al (2023) mScarlet3: a brilliant and fast-maturing red fluorescent protein. Nat Methods 20:541–545
16. Ng YK et al (2013) Expanding the repertoire of gene tools for precise manipulation of the

Clostridium difficile genome: allelic exchange using pyrE alleles. PLoS One 8:e56051

17. McAllister KN, Bouillaut L, Kahn JN, Self WT, Sorg JA (2017) Using CRISPR-Cas9-mediated genome editing to generate C. difficile mutants defective in selenoproteins synthesis. Sci Rep 7:14672
18. Maikova A, Kreis V, Boutserin A, Severinov K, Soutourina O (2019) Using an endogenous CRISPR-Cas system for genome editing in the human pathogen Clostridium difficile. Appl Environ Microbiol 85:e01416–e01419
19. Ticer TD et al (2024) Bacterial metabolites influence the autofluorescence of Clostridioides difficile. Front Microbiol 15:1459795
20. Coda S et al (2014) Fluorescence lifetime spectroscopy of tissue autofluorescence in normal and diseased colon measured ex vivo using a fiber-optic probe. Biomed Opt Express 5:515–538
21. Wang S et al (2025) Cell wall protein 2 as a vaccine candidate protects mice against Clostridioides difficile infection. Vaccine 13:21
22. Fagan RP, Fairweather NF (2011) *Clostridium difficile* has two parallel and essential sec secretion systems. J Biol Chem 286:27483–27493
23. Putnam EE, Nock AM, Lawley TD, Shen A (2013) SpoIVA and SipL are Clostridium difficile spore morphogenetic proteins. J Bacteriol 195:1214–1225
24. (2011) Guide for the care and use of laboratory animals: 8th edn. National Academies Press, Washington, DC. https://doi.org/10.17226/12910
25. Reeves AE et al (2011) The interplay between microbiome dynamics and pathogen dynamics in a murine model of Clostridium difficile infection. Gut Microbes 2:145–158
26. Trzilova D, Warren MAH, Gadda NC, Williams CL, Tamayo R (2022) Flagellum and toxin phase variation impacts intestinal colonization and disease development in a mouse model of Clostridioides difficile infection. Gut Microbes 14:2038854
27. Felts L, Kohli V, Marr JM, Schumacher J, Schlicker O (2020) An Introduction to Computational Clearing. https://www.leica-microsystems.com/science-lab/life-science/an-introduction-to-computational-clearing/
28. Chao Z, Xu W (2021) A new general maximum intensity projection technology via the hybrid of U-net and radial basis function neural network. J Digit Imaging 34:1264–1278
29. Taqi SA, Sami SA, Sami LB, Zaki SA (2018) A review of artifacts in histopathology. J Oral Maxillofac Pathol 22:279
30. Zhou H, Li J, Guan Y, He H, Huang Fu L (2022) Experimental study of different dehydration methods in the process of preparing frozen brain sections. Ibrain 10:164–171
31. Zhanmu O et al (2019) Maintenance of fluorescence during paraffin embedding of fluorescent protein-labeled specimens. Front Neurosci 13:752
32. Ren M et al (2018) Simultaneous acquisition of multicolor information from neural circuits in resin-embedded samples. Front Neurosci 12
33. Kirschnick N et al (2021) Rapid methods for the evaluation of fluorescent reporters in tissue clearing and the segmentation of large vascular structures. iScience 24:102650

Chapter 16

Clostridioides difficile Infection in Animals

Francisco A. Uzal

Abstract

Clostridioides difficile is the cause of the *C. difficile* infection (CDI) in humans and several other animal species. The microorganism is also found in the environment and in the gastrointestinal tract of healthy individuals of multiple animals of a variety of species. Although CDI in most domestic animals has historically been described in individuals receiving antibiotic treatment, or in the case of horse, hospitalized animals, the number of cases in which no antibiotic treatment or hospitalization association was known has been increasing over the past few years. Co-infections with other bacteria or parasites, intestinal displacements, transport, or surgical or medical treatment are also possible predisposing factors for CDI in horses. In most animals species, CDI is not age-associated. Although it has been suggested that CDI may be a zoonosis, this has not been demonstrated.

Key words Animals, *Clostridioides difficile*, *Clostridioides difficile* infection, One health

1 Introduction

Clostridioides difficile is responsible for *C. difficile*-associated infection (CDI) in humans and animals [1, 2]. While CDI in humans was traditionally an almost exclusively hospital-associated disease, recently there has been an increase of so-called community-associated cases, with no association to hospitalization [3]. A similar situation is observed in animals, particularly horses, where hospitalization does not seem critical any longer [4]. Likewise, although antibiotic treatment is still a significant predisposing factor for CDI in humans and animals, many cases occur in animals with no previous antibiotic treatment [4].

C. difficile may be present in multiple environments, including animal and human hospitals [4, 5]. Many healthy individuals of several animal species can carry the microorganism in the intestine, which is then shed to the environment, and becomes a source of infection for humans and other animals [6, 7]. Although there is growing evidence of the potential role of *C. difficile* as a zoonotic agent [8–11], no definitive proof of this role has been provided.

Kurt H. Piepenbrink and Jennifer M. Auchtung (eds.), *Clostridioides difficile: Methods and Protocols*, Methods in Molecular Biology, vol. 3046, https://doi.org/10.1007/978-1-0716-5328-9_16,

In this chapter we review and discuss the CDI in animals with particular emphasis on diagnosis and diagnostic criteria.

2 Horses

CDI can happen in horses of any age, as individual cases or, less commonly, outbreaks [6, 12, 13]. Although hospitalization and antibiotic therapy are still considered important predisposing factors for CDI in horses [12–14], cases in which neither of these factors are present can occur [4, 15]. Microbiota disturbance by factors such as transportation, surgery, concurrent infections, changes in feed, intestinal displacements, stress, and fasting have been also suggested, but not proven, to also predispose CDI [13, 14, 16, 17].

The clinical signs of CDI in horses may range from mild diarrhea to severe colic with hemorrhagic enterocolitis [12–14]. Sudden death may also occur, albeit rarely [12, 14].

While any part of the intestine may be affected, young foals tend to have lesions mostly in the small intestine, while older foals and adult horses tend to have lesions in the cecum and colon [12, 14]. Fibrino-necrotizing enteritis, enterocolitis, and colitis, in many cases with mucosal and submucosal thrombosis, are characteristic of the disease (Figs. 1, 2, and 3). Microscopically, there is mucosal necrosis (Fig. 4), with thrombosis (Fig. 5) and occasionally volcano lesions, characterized by the effusion of neutrophils through intestinal ulcers may be observed in some cases, and these lesions are highly suggestive of CDI (Fig. 6). Congestion and mild inflammation of the intestinal mucosa may be the only lesions observed in milder cases. A large number of bacteria may be observed in the intestinal mucosa (Fig. 7). While the infection is

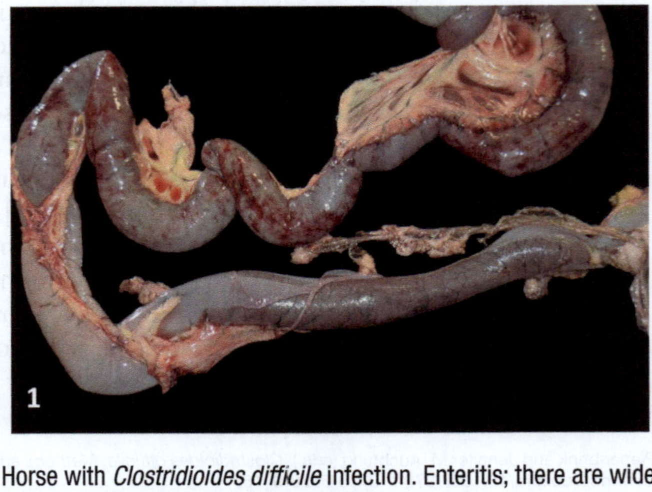

Fig. 1 Horse with *Clostridioides difficile* infection. Enteritis; there are widespread serosal hemorrhages. (Photo courtesy of J. Asin)

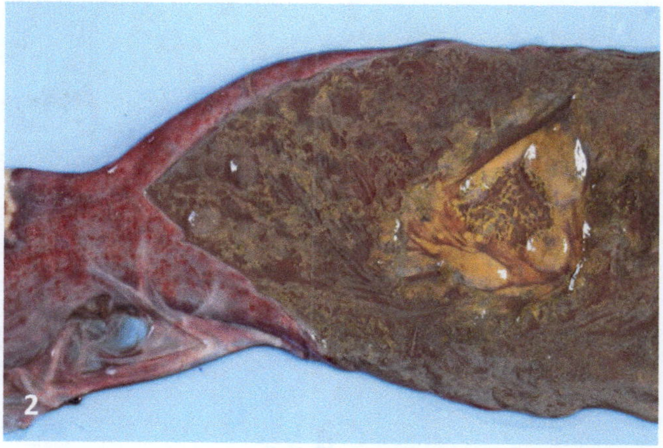

Fig. 2 Horse with *Clostridioides difficile* infection. Fibrino-necrotizing enteritis. The mucosa is covered by a fibrinous pseudo-membrane

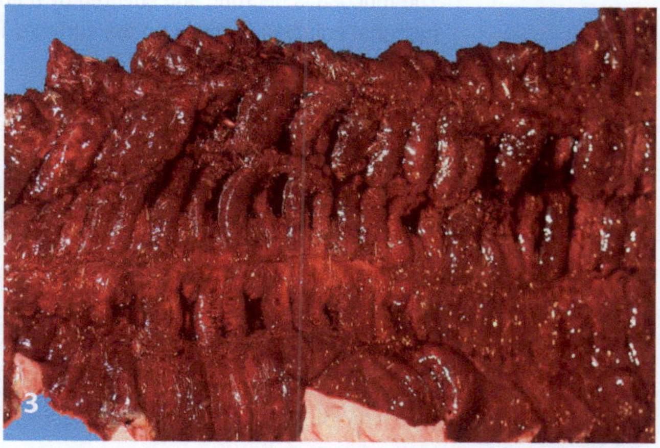

Fig. 3 Horse with *Clostridioides difficile* infection. Hemorrhagic colitis. The mucosa is diffusely congested and hemorrhagic

most commonly limited to the intestinal tract, cases in which septicemia occur have been also reported [12, 14, 18, 19].

3 Pigs

CDI in pigs is characterized by diarrhea and typhlocolitis. In this species the disease affects typically one-week-old piglets [4, 20]. Feces from the sow are the source of colonization for the piglets, which do not require antibiotic treatment to become infected. The infection occurs soon after birth via the fecal–oral route [21]. In severe outbreaks, the morbidity can be as high as 100%, although the mortality is usually lower [22]. Low growth rate, abdominal distention, ventral edema, and yellow diarrhea may

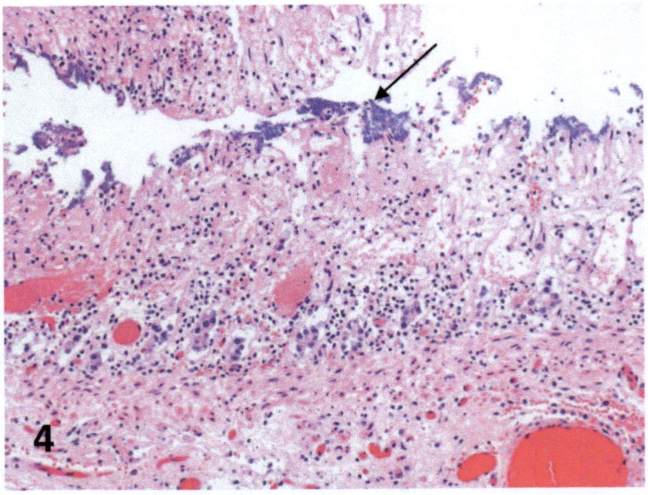

Fig. 4 Horse with *Clostridioides difficile* infection. Fibrino-necrotizing enteritis. The mucosa is necrotic and covered by a fibrinous pseudo membrane. There are bacterial colonies (arrows) on the surface. The mucosa is covered by a fibrinous pseudo membrane. HE

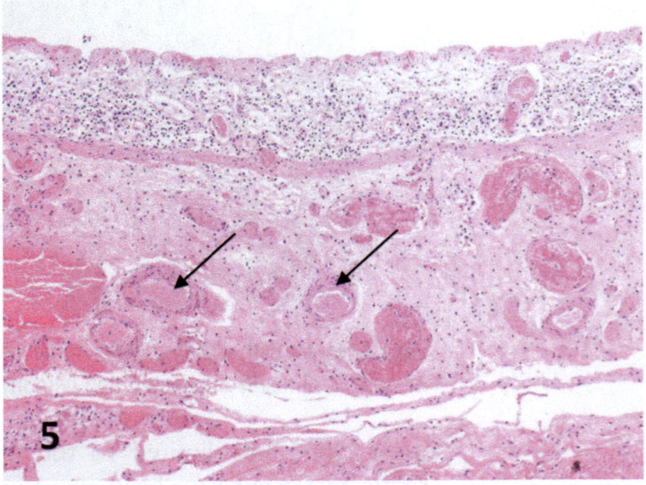

Fig. 5 Horse with *Clostridioides difficile* infection. Necrotizing colitis. There are multiple thrombi (arrows) in the mucosa and submucosa. HE

be observed in mild cases of CDI. Colic and hemorrhagic diarrhea may be observed in more severe cases [23]. As in horses, sudden death can occasionally occur [4, 23].

Gross lesions on post-mortem examination include distention of the colon with soft-to-watery contents and yellow contents. Pseudo-membranes covering the mucosa and mesocolonic edema are frequently observed (Figs. 8 and 9). Although not specific for CDI, the edema of the mesocolon is a characteristic lesion of the

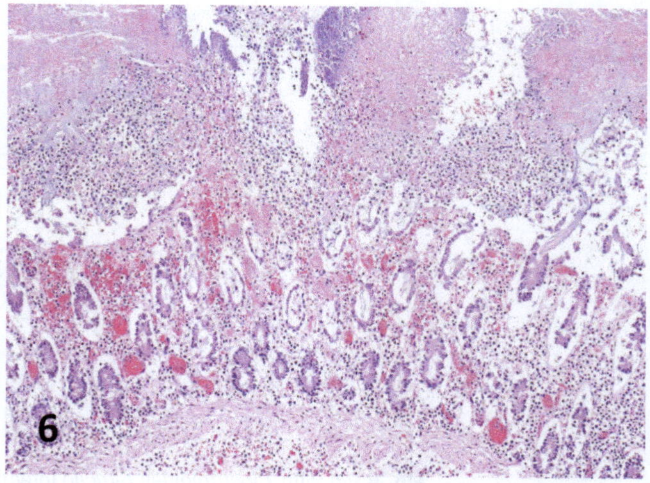

Fig. 6 Horse with *Clostridioides difficile* infection. Colitis with volcano lesions. Neutrophils are being exuded from the lamina propria into the lumen, through mucosal ulcers. HE

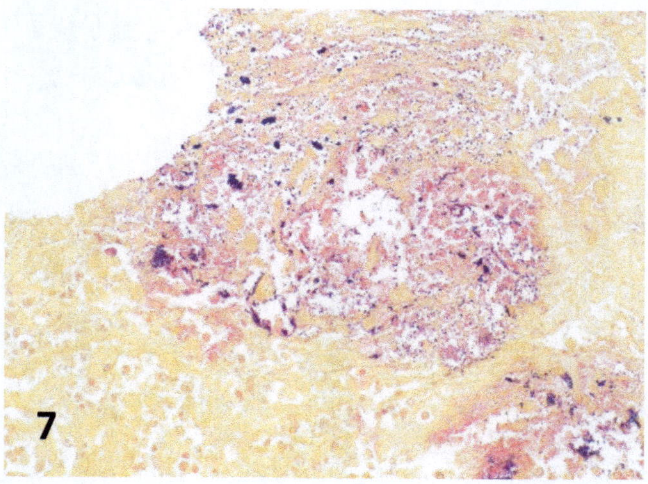

Fig. 7 Horse with *Clostridioides difficile* infection. Necrotizing colitis. There are myriad gram-positive bacilli and gram-negative bacilli in the necrotic mucosa. This horse had a *Clostridioides* difficile and *Salmonella* Typhimurium co-infection. Gram

condition, and its presence in neonatal piglets supports a diagnosis of CDI [4, 24].

Histologically, there is fibrinosuppurative colitis [4, 25, 26] and volcano lesions similar to those described for horses above are frequently seen (Fig. 10). Edema of the submucosa and mesocolon (Fig. 11) are almost constant lesions [4].

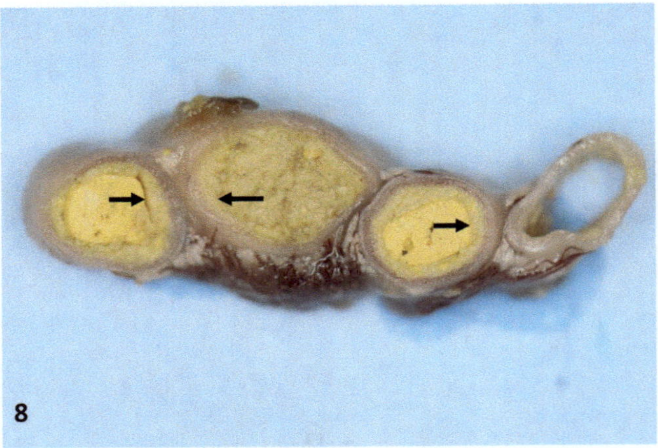

Fig. 8 Pig with *Clostridioides difficile* infection. Fibrino-necrotizing colitis. There is yellow content in the lumen and the mucosa is covered by a fibrinous pseudomembrane (arrows)

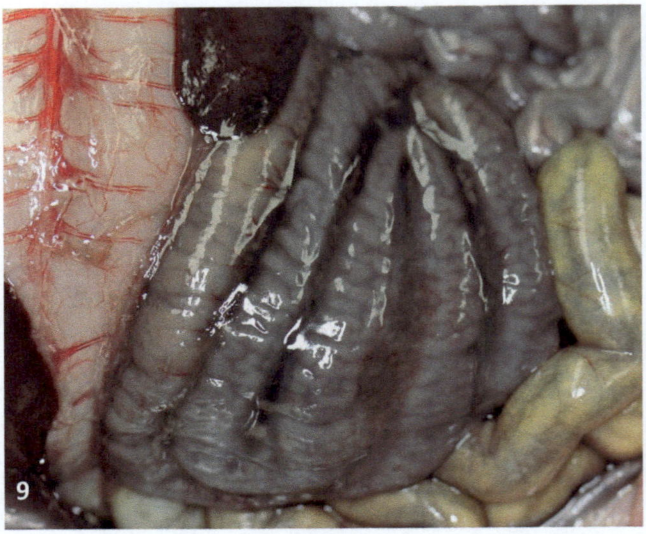

Fig. 9 Pig with *Clostridioides difficile* infection. Fibrino-necrotizing colitis. The spiral colon mesentery is expanded by edema

4 Rabbits

C. difficile causes enteric disease in rabbits [27]. The condition is, however, frequently associated with co-infection with other microorganisms [28]. In rabbits, the disease is age-dependent, with neonatal rabbits being resistant to CDI, which was postulated, but not proved, to be related to the lack of toxin receptors in the intestine [29]. Antibiotic treatment is the main predisposing factor for CDI in rabbits [30, 31] although cases have been reported in

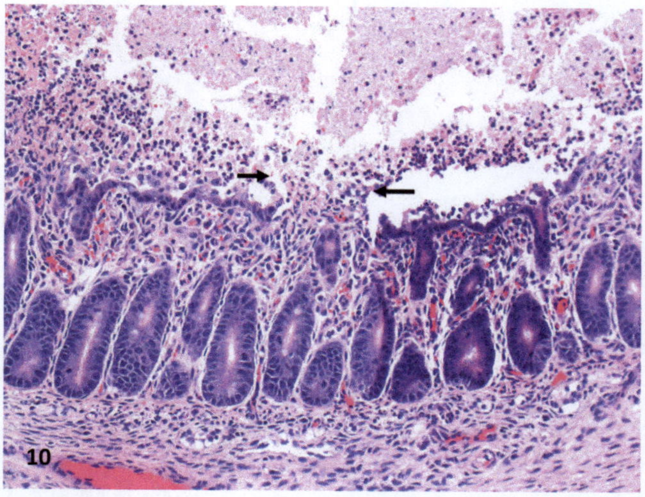

Fig. 10 Pig with *Clostridioides difficile* infection. Colitis with volcano lesions. Neutrophils are being exuded from the lamina propria into the lumen (arrows), through mucosal ulcers. HE

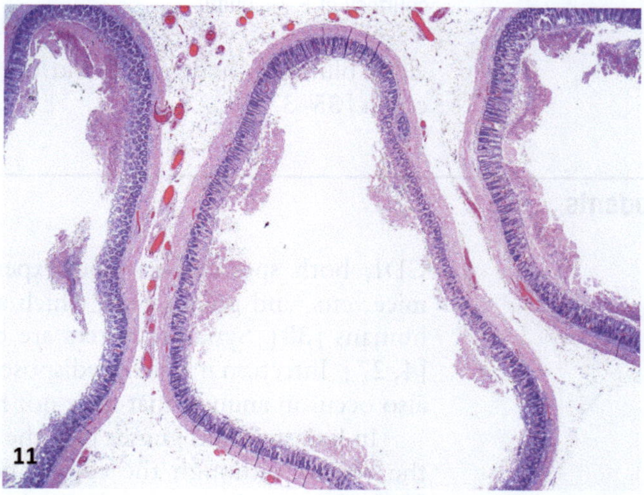

Fig. 11 Pig with *Clostridioides difficile* infection. Fibrino-necrotizing colitis. The mesentery is severely expanded by edema. HE

specific pathogen free rabbits without prior antibiotic treatment [32].

Clinical signs include anorexia, abdominal distention, and diarrhea, although sudden death can also occur [32–34]. At necropsy, the small intestine and colon are distended with brown to red fluid and gas, and the mucosa is covered by a pseudomembrane. Serosal hemorrhages and hyperemia are frequently seen (Fig. 12) [35]. Histologically, there is pseudomembranous enteritis and/or colitis, with villus blunting in the small intestine and mucosal ulceration,

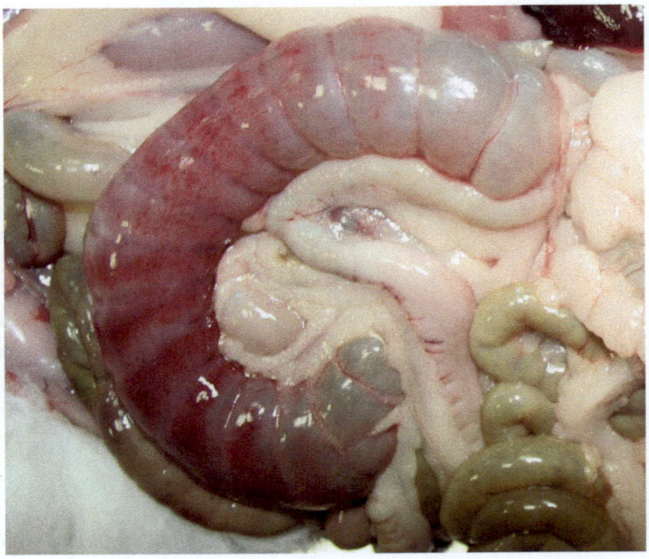

Fig. 12 Rabbit *Clostridioides difficile* infection. Enterocolitis. The large intestine is distended and the serosa is congested and has multiple hemorrhages. (Photo courtesy of F. Giannitti)

villus blunting, hyperemia, and edema in the small intestine and colon [35–37].

5 Rodents

CDI, both spontaneous and experimental, occurs in hamsters, mice, rats, and guinea pigs, which are used as models for CDI in humans [38]. Syrian hamsters are exquisitely susceptible to CDI [4, 27]. Infection may be predisposed by antibiotic therapy but can also occur in animals that have not received antibiotics [4, 31].

In hamsters and guinea pigs the lesions are usually seen only in the cecum, although the colon and ileum may also be affected [27]. Gross and microscopically the lesions are similar to those in other animal species [27].

6 Dogs and Cats

Although *C. difficile* has been found in the intestinal content and feces of a variable number of normal and diarrheic dogs and cats [15], CDI has not been reproduced to date in individuals of these species and the role of the microorganism in enteric disease of these species remains undetermined. Prevalence in healthy dogs ranges between 0% and 13.9% [15, 39, 40]. Same as in humans and horses, suggested risk factors for *C. difficile* shedding in dogs include antibiotic treatments and immunosuppressive drugs [41–45]. It was also suggested that shedding of *C. difficile* by dogs may be predisposed by antibiotic treatment [41] or immunocompromise [46] of the owners. Transmission from humans to dogs was also suggested although not proven [47].

C. difficile has been found in feces of healthy and diarrheic cats [48–52], although there is no clear evidence of association between this microorganism and enteric disease in this species.

Available data on *C. difficile* ribotypes present in cats is scant, but the little information available indicates that several strains associated with human disease, including 010, 106, 014/020, and ST42, can be found in cats [49, 52–54]. A potential zoonotic role of cats for CDI has been suggested but not proved [55].

7 Cattle

No final proof that *C. difficile* is associated with disease in cattle has been provided. Attempts to reproduce the disease experimentally in neonatal calves using the toxigenic strain ribotype 077, produced intestinal colonization but toxins were not detectable, and there were no signs of enteric disease [56]. Further studies are needed to investigate the relevance, if any, of *C. difficile* on intestinal disease in this species.

8 Sheep and Goats

As with cattle, there is no final evidence that *C. difficile* is associated with enteric disease in sheep or goats. Although *C. difficile* was detected in a small percentage of animals in several studies [57–60], none of the infected individuals had diarrhea or other significant sign of disease.

9 Poultry

While *C. difficile* has been isolated from fecal specimens obtained from both healthy and diarrheic birds, there is no compelling evidence to suggest that this microorganism is a cause of enteric disease in poultry [6].

10 Wildlife

There are sporadic reports of CDI in individual animals [61–63]; however, the significance of *C. difficile* in most wildlife species has not been thoroughly investigated [6].

11 Diagnosis of *Clostridioides difficile*-Associated Disease in Animals

Clinical history and signs, coupled with gross and microscopic postmortem findings, are helpful to establish a presumptive diagnosis of CDI [4]. Detection of *C. difficile* in intestinal contents or feces by culture or PCR in animals with diarrhea is supportive of a diagnosis of CDI, but since the microorganism can be found in a number of clinically healthy animals, this does not confirm the diagnosis [4]. Also, because non-toxigenic strains exist, this reduces further the value of detection of *C. difficile* in the intestinal content or feces of animals, unless isolation is followed by typing or the strains are directly detected by molecular methods that provide information on type. Detection of the organism is, however, very useful for epidemiological studies. Currently, most veterinary diagnostic laboratories confirm CDI by detection of toxins A and/or B in intestinal contents or feces by ELISA. Because small amounts of *C. difficile* toxins may be found in the intestinal content of clinically healthy animals, particularly piglets, it is therefore possible that toxin detection should not be considered the gold standard for CDI diagnosis. Additional research is needed to establish the correlation between toxin detection and disease in different animal species. In veterinary medicine, ribotyping is not routinely used for diagnostic purposes.

References

1. Keel K, Brazier JS, Post KW et al (2007) Prevalence of PCR ribotypes among *Clostridium difficile* isolates from pigs, calves, and other species. J Clin Microbiol 45:1963–1964
2. Uzal FA, Navarro MA, Li J et al (2018) Comparative pathogenesis of enteric clostridial infections in humans and animals. Anaerobe 53:11–20
3. Ofori E, Ramai D, Dhawan M et al (2018) Community-acquired *Clostridium difficile*: epidemiology, ribotype, risk factors, hospital and intensive care unit outcomes, and current and emerging therapies. J Hosp Infect 99:436–442
4. Uzal FA, Navarro MA, Asin J, et al (Submitted for publication) *Clostridioides difficile* infection in animals: a one health approach and literature review
5. Dharmasena M, Jiang X (2018) Isolation of toxigenic *Clostridium difficile* from animal manure and composts being used as biological soil amendments. Appl Environ Microbiol 84: e00738–e00718
6. Weese JS (2020) *Clostridium (Clostridioides) difficile* in animals. J Vet Diagn Invest 32: 213–221
7. Andrés-Lasheras S, Martín-Burriel I, Mainar-Jaime RC et al (2018) Preliminary studies on isolates of *Clostridium difficile* from dogs and exotic pets. BMC Vet Res 14:77
8. Alexiou S, Diakou A, Kachrimanidou M (2025) The role of *Clostridioides difficile* within the one health framework: a review. Microorganisms 13:429
9. Knetsch CW, Connor TR, Mutreja A et al (2014) Whole genome sequencing reveals potential spread of *Clostridium difficile* between humans and farm animals in The Netherlands, 2002 to 2011. Euro Surveill 19: 20954
10. Knight DR, Squire MM, Collins DA et al (2016) Genome analysis of *Clostridium difficile* PCR Ribotype 014 lineage in Australian pigs and humans reveals a diverse genetic repertoire and signatures of long-range interspecies transmission. Front Microbiol 7:2138

11. Knight DR, Riley TV (2019) Genomic delineation of zoonotic origins of *Clostridium difficile*. Front Public Health 7:164
12. Diab SS, Rodriguez-Bertos A, Uzal FA (2013) Pathology and diagnostic criteria of *Clostridium difficile* enteric infection in horses. Vet Pathol 50:1028–1036
13. Uzal FA, Navarro MA, Asin J et al (2022) Clostridial diseases of horses: a review. Vaccine 10:318
14. Diab SS, Songer G, Uzal FA (2013) *Clostridium difficile* infection in horses: a review. Vet Microbiol 167:42–49
15. Weese JS, Staempfli HR, Prescott JF (2001) A prospective study of the roles of *Clostridium difficile* and enterotoxigenic *Clostridium perfringens* in equine diarrhoea. Equine Vet J 33: 403–409
16. Uchida-Fujii E, Niwa H, Senoh M et al (2023) *Clostridioides difficile* infection in thoroughbred horses in Japan from 2010 to 2021. Sci Rep 13:13099
17. Nomura M, Kuroda T, Tamura N et al (2020) Mortality, clinical findings, predisposing factors and treatment of *Clostridioides difficile* colitis in Japanese thoroughbred racehorses. Vet Rec 187:e14
18. Mendonça FS, Navarro MA, Uzal FA (2022) The comparative pathology of enterocolitis caused by *Clostridium perfringens* type C, *Clostridioides difficile*, *Paeniclostridium sordellii*, *Salmonella enterica* subspecies *enterica* serovar Typhimurium, and nonsteroidal anti-inflammatory drugs in horses. J Vet Diagn Invest 34:412–420
19. Uzal FA, Diab SS, Blanchard P et al (2012) *Clostridium perfringens* type C and *Clostridium difficile* co-infection in foals. Vet Microbiol 156:395–402
20. Arruda PHE, Madson DM, Ramirez A et al (2013) Effect of age, dose and antibiotic therapy on the development of *Clostridium difficile* infection in neonatal piglets. Anaerobe 22: 104–110
21. Hopman NEM, Keessen EC, Harmanus C et al (2011) Acquisition of *Clostridium difficile* by piglets. Vet Microbiol 149:186–192
22. Songer JG, Jones R, Anderson MA et al (2007) Prevention of porcine *Clostridium difficile*-associated disease by competitive exclusion with nontoxigenic organisms. Vet Microbiol 124:358–361
23. Uzal FA, Navarro MA, Asin J et al (2023) Clostridial diarrheas in piglets: a review. Vet Microbiol 280:109691
24. Yaeger MJ, Kinyon JM, Glenn Songer J (2007) A prospective, case control study evaluating the association between *Clostridium difficile* toxins in the colon of neonatal swine and gross and microscopic lesions. J Vet Diagn Invest 19:52–59
25. Waters EH, Orr JP, Clark EG et al (1998) Typhlocolitis caused by *Clostridium difficile* in suckling piglets. J Vet Diagn Invest 10:104–108
26. McElroy MC, Hill M, Moloney G, et al (2015) Typhlocolitis associated with *Clostridium difficile* ribotypes 078 and 110 in neonatal piglets from a commercial Irish pig herd. Ir Vet J 69: 10
27. Keel MK, Songer JG (2006) The comparative pathology of *Clostridium difficile*-associated disease. Vet Pathol 43:225–240
28. Barthold SW, Griffey SM, Percy DH (2016) Rabbit. In: Pathology of laboratory rodents and rabbits. Wiley, Chichester, pp 253–324
29. Eglow R, Pothoulakis C, Itzkowitz S et al (1992) Diminished *Clostridium difficile* toxin a sensitivity in newborn rabbit ileum is associated with decreased toxin a receptor. J Clin Invest 90:822–829
30. Rehg JE, Lu YS (1981) *Clostridium difficile* colitis in a rabbit following antibiotic therapy for pasteurellosis. J Am Vet Med Assoc 179: 1296–1297
31. Rehg JE, Lu YS (1982) *Clostridium difficile* typhlitis in hamsters not associated with antibiotic therapy. J Am Vet Med Assoc 181:1422–1423
32. Perkins SE, Fox JG, Taylor NS et al (1995) Detection of *Clostridium difficile* toxins from the small intestine and cecum of rabbits with naturally acquired enterotoxemia. Lab Anim Sci 45:379–384
33. Taha MM, El-Helw HA, El-Sergany EF et al (2019) Identification of locally isolated *Clostridium difficile* from rabbits. WVJ 9:1–7
34. Carman RJ, Evans RH (1984) Experimental and spontaneous clostridial enteropathies of laboratory and free living lagomorphs. Lab Anim Sci 34:443–452
35. Diab SS, Uzal FA, Songer JG (2016) Diseases produced by *Clostridium difficile*. In: Clostridial diseases of animals. Wiley, Hoboken, pp 177–196
36. Warren CA, Calabrese GM, Li Y et al (2012) Effects of adenosine A_2A receptor activation and alanyl-glutamine in *Clostridium difficile* toxin-induced ileitis in rabbits and cecitis in mice. BMC Infect Dis 12:13
37. Lima AA, Lyerly DM, Wilkins TD et al (1988) Effects of *Clostridium difficile* toxins A and B in rabbit small and large intestine in vivo and on cultured cells in vitro. Infect Immun 56:582–588

38. Best EL, Freeman J, Wilcox MH (2012) Models for the study of *Clostridium difficile* infection. Gut Microbes 3:145–167
39. Struble AL, Tang YJ, Kass PH et al (1994) Fecal shedding of *Clostridium difficile* in dogs: a period prevalence survey in a veterinary medical teaching hospital. J Vet Diagn Invest 6:342–347
40. Finsterwalder SK, Loncaric I, Cabal A, et al (2022) Dogs as carriers of virulent and resistant genotypes of *Clostridioides difficile*. Zoonoses Public Health 69:673–681
41. Lefebvre P, Cariou B, Lien F et al (2009) Role of bile acids and bile acid receptors in metabolic regulation. Physiol Rev 89:147–191
42. Lefebvre SL, Reid-Smith R, Boerlin P et al (2008) Evaluation of the risks of shedding *Salmonellae* and other potential pathogens by therapy dogs fed raw diets in Ontario and Alberta. Zoonoses Public Health 55:470–480
43. Albuquerque C, Pagnossin D, Landsgaard K et al (2021) The duration of antibiotic treatment is associated with carriage of toxigenic and non-toxigenic strains of *Clostridioides difficile* in dogs. PLoS One 16:e0245949
44. Clooten JK, Kruth SA, Weese JS (2003) Genotypic and phenotypic characterization of *Clostridium perfringens* and *Clostridium difficile* in diarrheic and healthy dogs. J Vet Intern Med 17:123; author reply 123
45. Hussain I, Sharma RK, Borah P et al (2015) Isolation and characterization of *Clostridium difficile* from pet dogs in Assam, India. Anaerobe 36:9–13
46. Weese JS, Finley R, Reid-Smith RR et al (2010) Evaluation of *Clostridium difficile* in dogs and the household environment. Epidemiol Infect 138:1100–1104
47. Loo VG, Brassard P, Miller MA (2016) Household transmission of *Clostridium difficile* to family members and domestic pets. Infect Control Hosp Epidemiol 37:1342–1348
48. Clooten J, Kruth S, Arroyo L et al (2008) Prevalence and risk factors for *Clostridium difficile* colonization in dogs and cats hospitalized in an intensive care unit. Vet Microbiol 129: 209–214
49. Schneeberg A, Rupnik M, Neubauer H et al (2012) Prevalence and distribution of *Clostridium difficile* PCR ribotypes in cats and dogs from animal shelters in Thuringia, Germany. Anaerobe 18:484–488
50. Álvarez-Pérez S, Blanco JL, Harmanus C et al (2017) Prevalence and characteristics of *Clostridium perfringens* and *Clostridium difficile* in dogs and cats attended in diverse veterinary clinics from the Madrid region. Anaerobe 48: 47–55
51. Samir A, Abdel-Moein KA, Zaher HM (2021) Molecular detection of toxigenic *Clostridioides difficile* among diarrheic dogs and cats: a mounting public health concern. Vet Sci 8:88
52. Silva ROS, Ribeiro MG, de Paula CL et al (2020) Isolation of *Clostridium perfringens* and *Clostridioides difficile* in diarrheic and nondiarrheic cats. Anaerobe 62:102164
53. Rabold D, Espelage W, Abu Sin M et al (2018) The zoonotic potential of *Clostridium difficile* from small companion animals and their owners. PLoS One 13:e0193411
54. Silva ROS, Rupnik M, Diniz AN et al (2015) *Clostridium difficile* ribotypes in humans and animals in Brazil. Mem Inst Oswaldo Cruz 110:1062–1065
55. Garza MA, Thomas B, Saleh A et al (2023) Look what the cat dragged in! Recurrent *Clostridioides difficile* from a household cat. Am J Case Rep 24:e940923
56. Rodriguez-Palacios A, Stämpfli HR, Stalker M et al (2007) Natural and experimental infection of neonatal calves with *Clostridium difficile*. Vet Microbiol 124:166–172
57. Romano V, Albanese F, Dumontet S et al (2012) Prevalence and genotypic characterization of Clostridium difficile from ruminants in Switzerland. Zoonoses Public Health 59:545–548
58. Knight DR, Riley TV (2013) Prevalence of gastrointestinal *Clostridium difficile* carriage in Australian sheep and lambs. Appl Environ Microbiol 79:5689–5692
59. Avberšek J, Pirš T, Pate M et al (2014) *Clostridium difficile* in goats and sheep in Slovenia: characterisation of strains and evidence of age-related shedding. Anaerobe 28:163–167
60. Rahimi E, Jalali M, Weese JS (2014) Prevalence of *Clostridium difficile* in raw beef, cow, sheep, goat, camel and buffalo meat in Iran. BMC Public Health 14:119
61. Silva ROS, D'elia ML, de Magalhães Soares DF et al (2013) *Clostridium difficile*-associated diarrhea in an ocelot (*Leopardus pardalis*). Anaerobe 20:82–84
62. Orchard JL, Fekety R, Smith JR (1983) Antibiotic-associated colitis due to *Clostridium difficile* in a Kodiak bear. Am J Vet Res 44: 1547–1548
63. Anderson CE, Haulena M, Zabek E et al (2015) Clinical and epidemiologic considerations of *Clostridium difficile* in harbor seals (*Phoca vitulina*) at a marine mammal rehabilitation center. J Zoo Wildl Med 46:191–197

Chapter 17

Isolating *C. difficile* from Animal Samples

Deiziane V. S. Costa and Cirle A. Warren

Abstract

Clostridioides difficile (*C. difficile*), which has historically denominated *Bacillus difficile,* was first isolated by Hall and O'Toole in fecal samples of newborns during the first 10 days of life in 1935. Isolation of *C. difficile* from animal fecal specimens is critical for understanding the susceptibility and virulence of circulating strains through in vitro and preclinical studies, helping clinicians and researchers understand the different outcomes of CDI in the host associated with the pathogen's virulence. In this chapter, we will describe step-by-step procedures for isolating *C. difficile* from fecal specimens from animals with a focus on a method using taurocholate Cycloserine–Cefoxitin Fructose Agar (TCCFA) plates preceded by an alcohol shock.

Key words Bacterial isolation, Antibiotic selection, Clostridioides difficile

1 Introduction

Clostridioides difficile (*C. difficile*), which has historically denominated *Bacillus difficile,* was first isolated by Hall and O'Toole in fecal samples of newborns during the first 10 days of life [1]. Hall and O'Toole found that this new anaerobic bacterium could cause disease in guinea pigs and rabbits and questioned the association between *C. difficile* and unexplained bloody diarrhea and confusion in newborns [1]. Hall and O'Toole used a jar system and blood agar to isolate *C. difficile* from these samples, thereby obtaining successful isolation in 40% (4/10) of the infants' samples tested [1]. Later, the combination of the use of anaerobic chambers and selective media to isolate *C. difficile* improved the efficiency of isolating *C. difficile* in up to 99% of stools and other samples from humans and other animals.

Various selective media have been used to isolate *C. difficile* from fecal specimens (Table 1). Cycloserine–Cefoxitin Fructose Agar (CCFA) [2, 3] is widely used. This medium includes antibiotics that inhibit the growth of other bacteria (cycloserine and cefoxitin) and sugar fermented by *C. difficile* (fructose). CCFA

Kurt H. Piepenbrink and Jennifer M. Auchtung (eds.), *Clostridioides difficile: Methods and Protocols*, Methods in Molecular Biology, vol. 3046, https://doi.org/10.1007/978-1-0716-5328-9_17,

Table 1
Selective medium to isolate *C. difficile*

Medium	Formulation	Colony characteristic
ChromID™ *C. difficile* (CDIF) [11, 12]	Commercial available (bioMérieux)	Asymmetric shape and black-colored
CCFA	40 g proteose peptone, 5 g sodium phosphate dibasic, 1 g potassium phosphate monobasic, 2 g sodium chloride, 6 g fructose, 15 g agar, 9.0 mg neutral red, 500 mg cycloserine, and 15.6 mg cefoxitin	Flat, irregular, and gray or white
CCFA [7]	Same formulation described above but supplemented with 5% egg yolk	Flat, irregular, and gray or white
CCFA [3]	Supplemented with 5% defibrinated horse blood	Flat, irregular, and gray or white
TCCFA [5, 8]	CCFA supplemented with 0.1% taurocholate and 7% defibrinated horse blood	Flat, irregular, and gray or white
CDPA [10]	BHI (6 g), peptic digest of animal tissue (6 g), pancreatic digest of gelatin (14.5 g), NaCl (5 g), dextrose (3 g), anhydrous Na_2HPO_4 (2.5 g), sodium taurocholate (0.1%), D-cycloserine (300 mg) and cefoxitin (8.5 mg), X-Gal (150 mg), 0.5 mM glycine, 4-methylphenol (0.025%), defibrinated horse blood (6%), and agar (12 g)	Blue color
CDMN	*C. difficile* Oxoid suplemented with moxalactam and norfloxacin	–

with egg yolk, which promotes bacterial growth, was developed in 1979 by George and collaborators [4], who improved the efficacy of isolating *C. difficile* from fecal specimens presumptively by detecting colonies with distinct morphology and fluorescent properties. Some variations to improve spore germination, such as adding taurocholate [5, 6] or enriching with 5% egg yolk [7] or horse blood [5], have been reported. A previous study comparing CCFA media with and without horse blood and taurocholate [8] showed that supplementation of CCFA with these two components increased recovery of *C. difficile* to 100% [8]. *C. difficile* moxalactam–norfloxacin (CDMN) supplemented with 0.1% taurocholate has also been used before inoculation of colonies in CCFA [9].

Other selective media used to isolate *C. difficile* from fecal specimens have also been reported. For instance, the *C. difficile* plate assay (CDPA) medium was previously developed to isolate *C. difficile* strains that produce toxins, which form blue colonies [10]. Another one is chromID™ *C. difficile* (CDIF), with *C. difficile* colonies being asymmetric and black [11, 12].

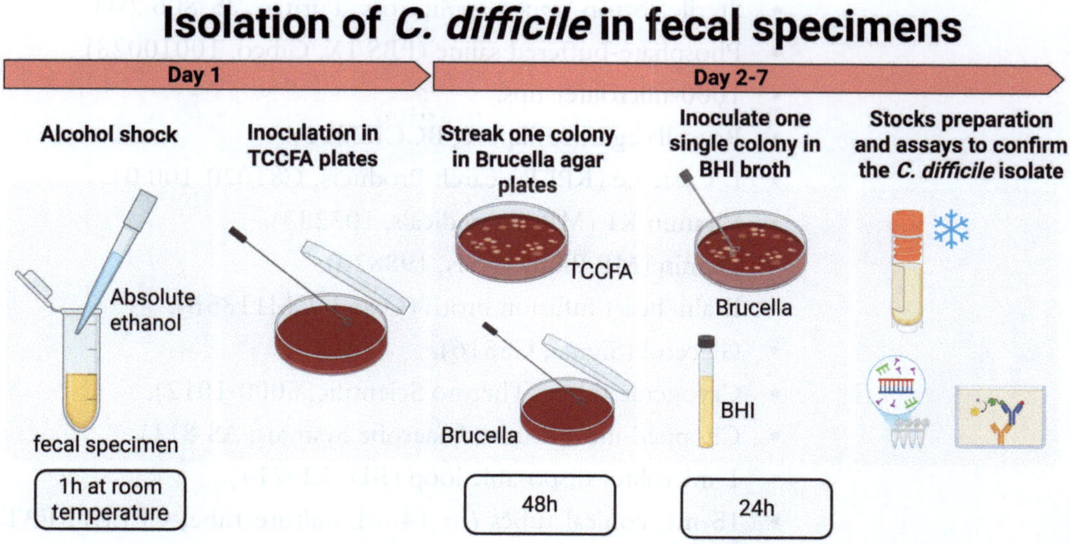

Fig. 1 Overall procedure to isolate *C. difficile* from fecal specimens of animals. (Figure created using BioRender)

Another source of variability in the method for *C. difficile* isolation is the agent used to kill non-*C. difficile* species prior to inoculation on selective medium, such as treatment of fecal specimens with absolute alcohol [5] or 3% chloroform [2].

Isolation of *C. difficile* from animal fecal specimens is essential for understanding the susceptibility and virulence of circulating strains through in vitro and preclinical studies, helping clinicians and researchers understand the different outcomes of CDI in the host associated with the pathogen's virulence. In this chapter, we will describe step-by-step procedures for isolating *C. difficile* from fecal specimens from animals with a focus on a method using TCCFA plates preceded by an alcohol shock (Fig. 1).

2 Materials

- Anaerobic chamber (10% CO_2, 10% H_2; and 80% N_2) or jar system.
- Taurocholate (Sigma, 86339).
- *Clostridium difficile* agar base (Oxoid, CM0601).
- Defibrinated horse blood (LAMPIRE biological laboratories, 7233401500ML).
- 2-mL microtubes.
- *C. difficile* selective supplement (Oxoid, SR0096E).
- Absolute ethanol (200 proof, Decon Laboratories).

- Sterile cotton-tipped applicators (Puritan, 25-806 2WC).
- Phosphate-buffered saline (PBS 1X, Gibco, 10010023).
- 1000-microliter tips.
- Brucella agar (Millipore, BCCK2392).
- L-Cysteine (RPI Research Products, C81020-100.0).
- Vitamin K1 (MP Biomedicals, 103283).
- Hemin (MP Biomedicals, 198820).
- Brain–heart infusion broth (Oxoid, CM1135B).
- Glycerol (Sigma, G5516).
- Cryogenic tubes (Thermo Scientific, 5000-1012).
- Chopped meat broth (Anaerobe Systems, AS-811).
- 1-microliter disposable loop (BD, 220214).
- 15-mL conical tubes (or 14-mL culture tube, CELLTREAT, 230439).
- 0.22 syringe filter (Millipore, SLGPM33RS).

3 Methods

3.1 Media Preparation

Schematics showing detailed methods for preparing TCCFA plates (Fig. 2), Brucella agar (Fig. 3), and supplemented BHI broth (Fig. 4) are depicted below.

- TCCFA plates.
- Brucella agar with supplements.
- BHI broth with supplements (BHIS).

3.2 Procedure

In preparation for the isolation procedure, add sufficient supplies or liquid media, if required, into the anaerobic chamber to allow sufficient time for the materials to be reduced in oxygen the day before the procedure.

- Day 1: Inoculation in TCCFA plates

1. Clean the anaerobic chamber, and then place the TCCFA plates in it (1 h before starting the next step).
2. Clean the hood with 10% bleach and add all the necessary materials to the hood: absolute alcohol and 1000 microliter tips.
3. Label 2-mL microtubes with the sample name.
4. Alcohol shock: Add 1 mL of fecal specimens to 1 mL of absolute ethanol in the labeled microtubes and incubate for 1 h at room temperature (*see* **Note 1**).

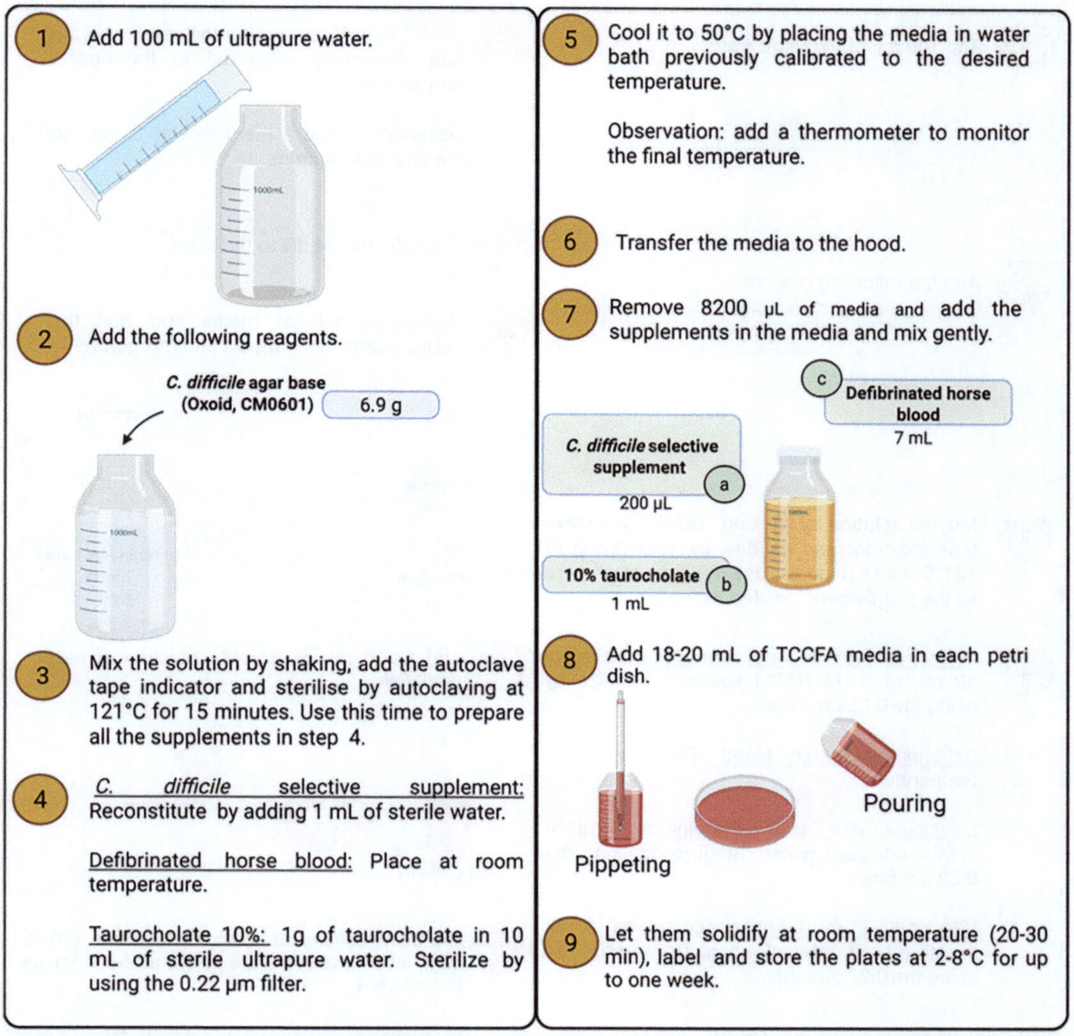

Fig. 2 Preparing TCCFA plates

5. Centrifuge at 3500 rpm for 10 min.
6. Remove the supernatant.
7. Place the microtubes containing the pellets in the anaerobic chamber.
8. Resuspend the pellets in 100 μL of sterile 1x PBS.
9. Inoculate 10 μL of the pellets into the TCCFA plates by using an inoculation loop (*see* **Note 2**).
10. Incubate the plates at 37 °C in anaerobic conditions.
11. From day 2 to 7, observe the growth of colonies on the TCCFA plates.

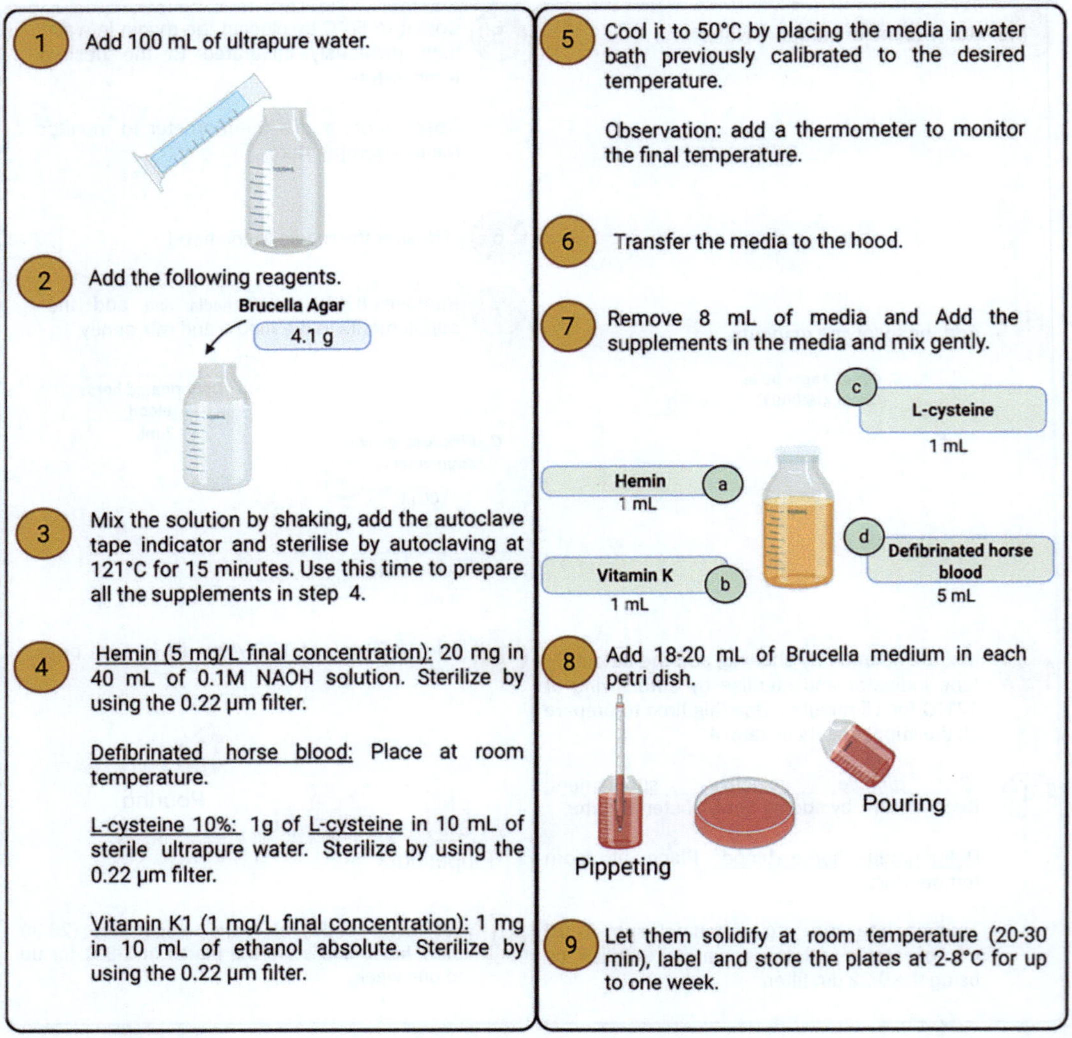

Fig. 3 Preparing TCCFA plates

12. Follow the steps below if you observe typical *C. difficile* colonies, which are characterized by being flat, irregular, and gray or white in TCCFA plates (Fig. 5).
13. Place the Brucella agar plates in the anaerobic chamber (2 h before starting the next step).
14. By using a 1-µL inoculation loop, transfer one *C. difficile* colony from the TCCFA plate to the Brucella agar plate and streak (zig-zag motion) the colony.
15. Incubate the plates at 37 °C in anaerobic conditions for 48 h.
16. Twenty-four hours after inoculating the colonies in Brucella agar plates, place BHIS media in the anaerobic chamber (with the lid partially open to allow removal of oxygen).

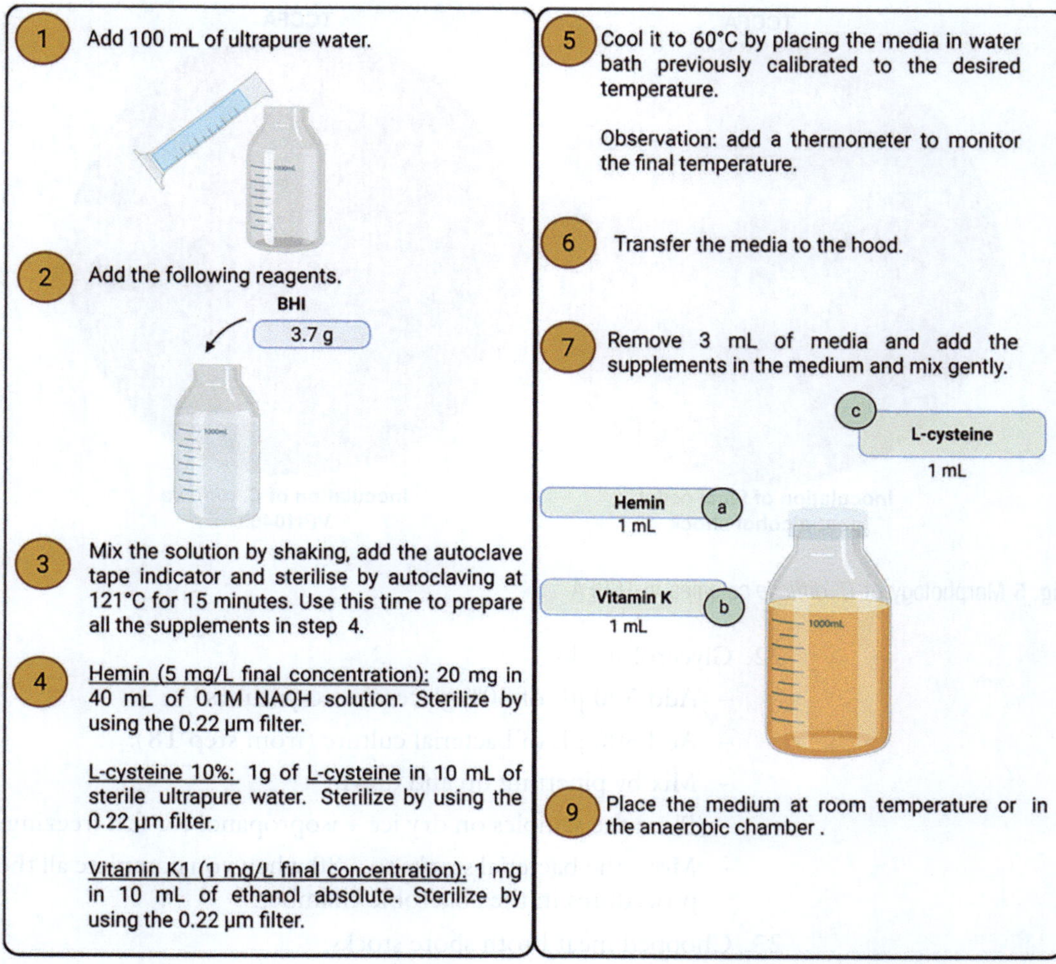

Fig. 4 Preparing BHI broth with supplements (BHIS)

17. Add 5 mL of BHIS media to a 15-mL conical tube (or 14-mL culture tube).
18. Transfer one colony of *C. difficile* from Brucella agar into the conical tube containing BHIS and vortex.
19. Incubate the bacteria at 37 °C in anaerobic condition for 24 h.
20. Place 50% glycerol solution (with the lid partially open to allow removal of oxygen), cryogenic tubes, and chopped meat broth in the anaerobic chamber 24 h before the start of step 22 (*see* **Note 3**).
21. Just before starting step 22, place a little container with dry ice and isopropanol in the anaerobic chamber for preparing the glycerol stocks.

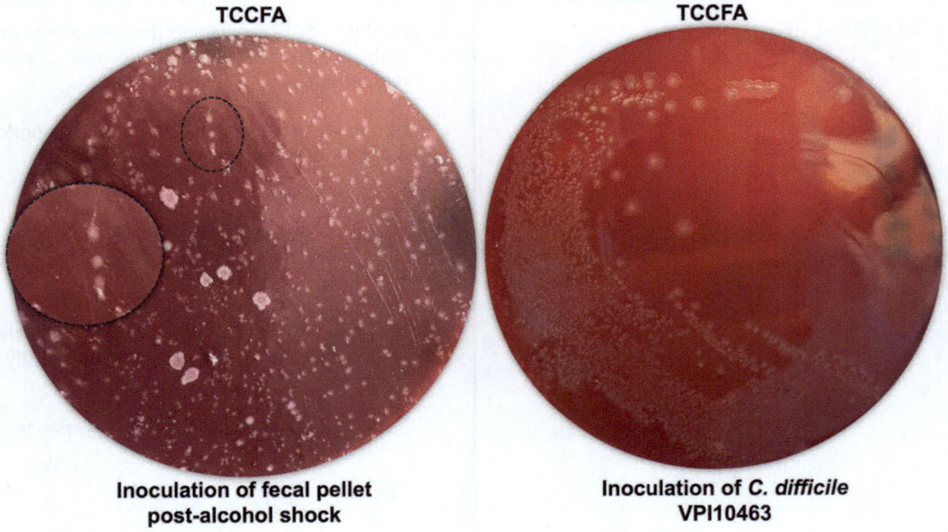

Fig. 5 Morphology of *C. difficile* colonies in TCCFA

22. Glycerol stocks:
    - Add 500 μL of 50% glycerol to cryotubes.
    - Add 500 μL of bacterial culture (from **step 18**).
    - Mix by pipetting up and down.
    - Place the samples on dry ice + isopropanol for fast freezing.
    - Move the bacterial stocks to −80 when you complete all the procedures in the anaerobic chamber.
23. Chopped meat broth spore stocks:
    - Add 100 μL of bacterial culture to chopped meat broth tubes.
    - Grow at 37 °C in anaerobic condition for 7 days.
    - Store the vials at room temperature, protected from light.
24. To confirm isolates as positive for *C. difficile*, several approaches can be used:
    - qPCR to detect the genes *tpi*, *tcdA*, *tcdB*, *cdtA,* and *cdtB*: 1 mL of bacterial culture for DNA extraction, as previously described [5].
    - ELISA: detection of *C. difficile* toxins in the supernatant of bacterial culture by using commercial kits available.
    - Aerotolerance: Inoculate 10 μL of bacterial culture in BHIS agar and incubate at 37 °C under aerobic condition for 24 h.
    - Matrix-assisted laser desorption ionization-time of flight (MALDI-TOF) mass spectrometry.
    - Gram stain.

4 Notes

1. Alternatively, you can add 500 μL of fecal specimens to 500 μL of absolute ethanol.
2. The pellet resuspended in PBS can be reused if the isolation is unsuccessful. In that case, store the pellet suspension at 2–8 °C.
3. To make 50% glycerol solution: 5 mL of 100% glycerol +5 mL of sterile water. Sterilize with a 0.22-μm filter.

References

1. Hall IC, O'Toole E (1935) Intestinal flora in new-borin infants - with a description of a new pathogenic anaerobe, Bacillus difficilis. Am J Dis Child 49(2):390–402. https://doi.org/10.1001/archpedi.1935.01970020105010
2. Ahn SW, Lee SH, Kim UJ, Jang HC, Choi HJ, Choy HE et al (2021) Genomic characterization of nine Clostridioides difficile strains isolated from Korean patients with Clostridioides difficile infection. Gut Pathog 13(1): 55. https://doi.org/10.1186/s13099-021-00451-3
3. Shirinda H, Smith AM, Prinsloo B, Kock MM, Moodley M, Said M et al (2025) Clostridioides difficile hypervirulent strain ST1 isolated from clinical stool specimens obtained from three Provinces in South Africa. Anaerobe 91: 102926. https://doi.org/10.1016/j.anaerobe.2024.102926
4. George WL, Sutter VL, Citron D, Finegold SM (1979) Selective and differential medium for isolation of Clostridium difficile. J Clin Microbiol 9(2):214–219. https://doi.org/10.1128/jcm.9.2.214-219.1979
5. Costa DVS, Pham NVS, Hays RA, Bolick DT, Goldbeck SM, Poulter MD et al (2022) Influence of binary toxin gene detection and decreased susceptibility to antibiotics among Clostridioides difficile strains on disease severity: a single-center study. Antimicrob Agents Chemother 66(8):e0048922. https://doi.org/10.1128/aac.00489-22
6. Wilson KH, Kennedy MJ, Fekety FR (1982) Use of sodium taurocholate to enhance spore recovery on a medium selective for Clostridium difficile. J Clin Microbiol 15(3):443–446. https://doi.org/10.1128/jcm.15.3.443-446.1982
7. Liao F, Li W, Gu W, Zhang W, Liu X, Fu X et al (2018) A retrospective study of community-acquired Clostridium difficile infection in Southwest China. Sci Rep 8(1):3992. https://doi.org/10.1038/s41598-018-21762-7
8. Tyrrell KL, Citron DM, Leoncio ES, Merriam CV, Goldstein EJ (2013) Evaluation of cycloserine-cefoxitin fructose agar (CCFA), CCFA with horse blood and taurocholate, and cycloserine-cefoxitin mannitol broth with taurocholate and lysozyme for recovery of Clostridium difficile isolates from fecal samples. J Clin Microbiol 51(9):3094–3096. https://doi.org/10.1128/JCM.00879-13
9. Tian TT, Zhao JH, Yang J, Qiang CX, Li ZR, Chen J et al (2016) Molecular characterization of Clostridium difficile isolates from human subjects and the environment. PLoS One 11(3):e0151964. https://doi.org/10.1371/journal.pone.0151964
10. Darkoh C, Dupont HL, Kaplan HB (2011) Novel one-step method for detection and isolation of active-toxin-producing Clostridium difficile strains directly from stool samples. J Clin Microbiol 49(12):4219–4224. https://doi.org/10.1128/JCM.01033-11
11. Rohana H, Azrad M, Nitzan O, Adler A, Binyamin D, Koren O et al (2020) Characterization of Clostridioides difficile strains, the disease severity, and the microbial changes they induce. J Clin Med 9(12). https://doi.org/10.3390/jcm9124099
12. Schwartz O, Rohana H, Azrad M, Shor A, Rainy N, Maor Y et al (2023) Characterization of community-acquired Clostridioides difficile strains in Israel, 2020–2022. Front Microbiol 14:1323257. https://doi.org/10.3389/fmicb.2023.1323257

INDEX

Kurt H. Piepenbrink and Jennifer M. Auchtung (eds.), *Clostridioides difficile: Methods and Protocols*, Methods in Molecular Biology, vol. 3046, https://doi.org/10.1007/978-1-0716-5328-9,

MIX
Papier aus verantwortungsvollen Quellen
Paper from responsible sources
FSC® C105338

If you have any concerns about our products,
you can contact us on
ProductSafety@springernature.com

In case Publisher is established outside the EU,
the EU authorized representative is:
Springer Nature Customer Service Center GmbH
Europaplatz 3, 69115 Heidelberg, Germany

Printed by Libri Plureos GmbH
in Hamburg, Germany